Double Envelope Houses

A Survey of Principles and Practice

William A. Shurcliff

Brick House Publishing Company
Andover, Massachusetts

R 8 AGED R7 INSULATION

BRAND NAME THERMAX

SOLD BY BORDWELL INSULATION

Published by Brick House Publishing Co., Inc.
34 Essex Street
Andover, Massachusetts 01810

Cover design: Ned Williams
Typesetting: Janet E. Lendall

Printed in the United States of America

Library of Congress Cataloging in Publication Data

Shurcliff, William A
Superinsulated houses and double-envelope houses.

Bibliography: p.
Includes index
1. Dwellings--Insulation. 2. Dwellings--Heating and ventilation. I. Title.
TH1715.S52 693.8'32 80-27198
ISBN 0-931790-19-0
ISBN 0-931790-18-2 (pbk.)

TABLE OF CONTENTS

Chapter 1
INTRODUCTION

TWO EXPLOSIVE GROWTHS

Two explosive growths are occurring: growth of superinsulated houses and growth of double-envelope houses. Estimates made by friends suggest that by August 1980 as many as 50 or 150 superinsulated houses had been completed and about 500 to 1500 others were under construction. The estimates for double-envelope houses were similar.

AUTHOR'S GOAL

My goal has been to describe the two new fields in general terms, compare them, describe specific houses in detail, explain the main principles used, raise questions as to the merits of individual design features, indicate some weak points, suggest possible improvements, and present lists of the main organizations and individuals involved.

DEFINITION OF SUPERINSULATED HOUSE

I use this term to mean a house that is situated in a cold climate (4000 degree days or colder) and:

receives only a modest amount of solar energy (for example, has a south-facing window area not exceeding 8% of the floor area)

is so well insulated, and is so airtight, that, throughout most of the winter, it is kept warm solely by (a) the modest amount of solar energy received through the windows and (b) miscellaneous within-house heat sources (intrinsic heat sources). Little auxiliary heat is needed: say less than 15% as much as is required by typical houses that were built before 1974 and are of comparable size.

By "miscellaneous heat sources within the house" I mean these intrinsic internal sources:

Stoves for cooking
Domestic hot water systems
Clothes dryers
Electric lights
Clothes washing machines
Dish washing machines
Human bodies
TV and radio sets
Refrigerators
Other electric devices (in kitchen, living areas, workshop, etc.)

By "auxiliary heat" I mean explicit auxiliary heat, e.g., heat from a system installed mainly for the purpose of space heating. Examples: oil or gas furnace, electric space heating system, heat-pump, wood-burning stove in living room.

The 8% limit on south window area was chosen because, if the area is much greater, intake of solar energy may be more important than conserving heat from miscellaneous internal sources. In other words, the house may be more a solar house than a superinsulated house -- because large windows, even when double-glazed, lose large amounts of heat at night. See Ref. S-235ee.

The 15% limit on auxiliary heat (15% relative to comparable-size houses in 1974) was chosen because a house that conforms to this limit can get through the winter half-way tolerably (or better) even if auxiliary heat is cut off entirely. Specifically, the house will never cool down to 32°F; pipes will never freeze; the pipes will never have to be drained; faucets, sinks, toilets, tubs, etc., will continue to operate normally. Furthermore, it would take only a little additional heat, as from a wood stove or one or two portable electric heaters, to keep such a house fairly comfortable even when fuel supplies are cut off.

Another reason for adopting this limit is to shorten the list of houses that qualify, i.e., to focus attention on the few most interesting of the very-well-insulated houses. Only a few of today's passively solar heated houses qualify. Almost no actively solar heated houses qualify.

DEFINITION OF A DOUBLE-ENVELOPE HOUSE

I use double-envelope house to mean a house that:

has, throughout about half or more of the combined area of roof (or attic) and north wall, two envelopes with a continuous airspace at least 6 in. thick between them,

has a large greenhouse integral with the south side,

has a large below-grade airspace, such as a crawl space, or a basement, or sub-basement space,

has full continuity from one airspace to the next, so that air is free to circulate around a complete circuit, or convective loop, that includes the greenhouse, the roof or attic, the north wall system, and the crawl space or equivalent.

I avoid making a performance definition. Not enough is known about the performance of such houses.

NOW THERE ARE SIX TYPES OF SOLAR-HEATED HOUSES

The four well-known types of solar-heated houses are:

Houses with active solar heating systems

Houses with combined active and passive solar heating systems

Houses with direct-gain passive solar heating

Houses with indirect-gain passive solar heating.

The time has come to add two types to the list:

Houses that are superinsulated (and use only a moderate amount of solar heating -- passive solar heating)

Houses that employ double-envelopes in conjunction with a large integral greenhouse along the south side.

MY TENTATIVE CONCLUSION

Both types of houses described here appear to have enthusiastic followers. The enthusiasm is shared by the designers, builders, and occupants. Both types of houses require very little auxiliary heat.

The double-envelope houses seem to especially delight persons who like greenhouses. Partly because I have never had any real familiarity with greenhouses and partly because the design (and operation) of double-envelope houses is beset with many not-yet-answered questions as to how and how well various components and systems perform, I do not have a full enthusiasm for such houses. I retain a general skepticism concerning certain specific features, which I discuss at length (too great length?) in later chapters. Perhaps some of the performance data that will be amassed in the next year or two will dispell some of the skepticism.

Contrariwise, I am wholeheartedly enthusiastic about super-insulated houses. They are simple and inexpensive, and they seem fully understandable and fully successful. (Sometimes I feel that they are so very simple and understandable as to be no fun! Surely mystery and complexity add zest to our lives?)

Neither type of house has been adequately tested under a wide range of conditions. Performance data are still very fragmentary. It is too soon to make any firm judgments. We shall know a lot more in another year or two. Meanwhile a great debt of gratitude is owed to the pioneers who have pushed these exciting developments forward, using their own inventiveness and drive and their own money.

ADDITIONAL INFORMATION WELCOMED

I will be much obliged to persons who supply me with additional information such as would help me correct errors and omissions in this book or would help me in preparing a later edition. Such information should be sent to William A. Shurcliff, 19 Appleton Street, Cambridge, MA 02138.

ACKNOWLEDGEMENT

I am indebted to many pioneers in these fields for explaining to me what they have done and why they did it. They have explained the design principles used, the specific architectural features used, and the feelings of the occupants of the houses in question.

I especially wish to thank Professor Wayne L. Shick of the Small Homes Council of the University of Illinois, Eugene H. Leger of Vista Homes, Lee P. Butler of Ekose'a, Tom Smith of Positive Technologies Corporation, David A. Robinson of Mid-American Solar Energy Complex, Robert S. Dumont of the National Research Council of Canada. Much help was received also from Don Booth of Community Builders, Robert Mastin, N.B. Saunders, G.S. Dutt, R.O. Smith, Hank Huber.

Photographic credits: E.H. Leger (Leger House), Robert Mastin (Mastin House), Johs. Gunnarshaug (Norwegian House at Aas).

WARNINGS

Much of the information presented here has not been fully confirmed. Serious errors of fact, or errors in attributing praise or blame, may occur. No reliance should be placed on any specific statement without making independent confirmation.

Some of the devices, systems, etc., described may be covered by patents, and some of the names and designs may be covered by copyright or trademark. I have made no systematic attempt to learn about, or present information on, patents, copyrights, trademarks.

Btu-to-kWh CONVERSION

1 Btu = 0.0002931 kWh. 1 kWh = 3412 Btu.

LEGER HOUSE

A superinsulated house. View of south face. For details see Chapter 3.

View of northwest corner.

Views of construction near Southeast corner

MASTIN HOUSE

A double-envelope house. View of south face. For details see Chapter 8.

North face

NORWEGIAN HOUSE

At Aas, south face. See Chapter 11.

Chapter 2
SUPERINSULATED HOUSES: PRELIMINARY CONSIDERATIONS

PRINCIPLE OF OPERATION OF A SUPERINSULATED HOUSE

The principle is easily stated: The house is so superbly insulated and so airtight that, throughout the winter, it is kept warm enough (or almost warm enough) by intrinsic sources of heat such as human bodies, light bulbs, cooking stove, etc., and by a modest amount of direct-gain passive solar heating. Very little auxiliary heat is needed: so little that there is no need for a furnace.

ALTERNATIVE NAMES

Other names might be used, but each has drawbacks. _Micro-load_ is not sufficiently restrictive; nearly all houses near the equator have only micro heat-loads. _Autonomous_ is a rare word, hard to spell; it has political connotations. _Mini-need_, like micro-load, would apply to most houses in the tropics. _Energy conserving_ is too bland; most of the buildings now being designed today are, in some sense, energy conserving. _Self-sufficient_ comes close, but is not specific enough; self-sufficient in what respect? Anyway, a house has no _self_.

Superinsulated is not an ideal name, but is good enough.

HALLMARKS OF A SUPERINSULATED HOUSE

The main distinguishing characteristics, or hallmarks, of a superinsulated house are:

Truly superb insulation. Even at the sills, headers, eaves, window frames, door frames, and electric outlet boxes at least a moderate amount of insulation is provided. In summary, the insulation is thick and thorough.

An almost airtight envelope. Even on windy days the rate of air change is low.

No added thermal mass. No Trombe wall, no water-filled drums, no concrete floors (except basement floor).

No very large south-window area. The amount of direct-gain passive solar heating is modest.

The combined area of windows on west, north, and east is less than the area of south windows. Perhaps only half.

No furnace. (There may be some electric heaters or a wood-burning stove, but these are used only rarely.)

No large system for distributing heat among the rooms.

No distorted shape of house.

Some corollaries:

There is no big added expense. The costs of the extra insulation and extra care in construction are largely offset by the savings from having no huge area of Thermopane, no huge expensive thermal shutters for huge south windows, no furnace, no big heat distribution system.

The passive solar heating system is almost incidental.

Room humidity remains at least fairly high throughout the winter. Sometimes it may be too high! There is no need for humidifiers (but there may be a need for a dehumidifier or an air-to-air heat-exchanger).

In summer the house usually stays cool automatically, if windows are opened wide each night. Even the south rooms usually stay cool: the area of south windows is small and, in summer, these windows are shaded by the wide eaves.

Because a superinsulated house does not need a great deal of solar radiation, it can perform well in urban sites and in moderately wooded areas. (Pointed out to me by G.S. Dutt).

INTRINSIC HEAT

I use intrinsic heat to mean that heat produced (in living room, kitchen, bedrooms, basement rooms, etc.) by (a) people, (b) miscellaneous sources such as electric light bulbs, cooking stove, etc.

Not included is heat from furnace, stove used just for space heating, electric space heaters, or the heat that is important to domestic hot water and is lost when the water goes down the drain. Also I exclude solar energy received via windows, via a greenhouse, or other collector.

Here I list some of the main sources of intrinsic heat, and I include estimates made by Massdesign Architects and Planners in its 1975 book "Solar Heated Houses for New England". See Bibliography item M-82a.

Source of intrinsic heat	Amount of intrinsic heat (Btu) produced in a typical house in a typical 24-hr midwinter day
Human bodies (two adults and children)	29,000
Cooking stove, microwave oven	18,600
Heater for DHW	4,100
Clothes dryer	8,300
Miscellaneous electric equipment: clothes washer, dish washer, refrigerator, TV and radio, hair dryer, blankets, toaster, coffee pot, blowers, fans	59,500
	119,500

Obviously, the estimate is a very rough one. From house to house there is, certainly, much variability.

Other Estimates

The designers of the Lo-Cal house have used the value 51,000 Btu/day; see Circular C2.3 (Bibl. S-184-C2.3). D. Lewis and W. Fuller have used the value 60,000 Btu/day; see the December 1979 *Solar Age*, p. 31. Some persons use this figure: 20,000 Btu per person per day.

Warning

As lower-power appliances come into use and families learn to cut down on the use of electrical-power-consuming equipment, the amount of intrinsic heat available will decrease. (Perhaps there will be a slight counter-tendency: perhaps more people will crowd into a given amount of floor-area, to reduce the per capita cost of rent and heat; if so, the amount of body heat contributed to the room air will increase.) Of course, cutting down on electrical appliance use in summer is helpful in every way.

ORIGIN OF THE SUPERINSULATED HOUSE

In America the first group to undertake large-scale exploration, development, and promotion of the superinsulated house was the Small Homes Council of the University of Illinois. Its efforts culminated in 1976 with the design of the Lo-Cal House.

The Lo-Cal House is a conceptual design: a carefully worked out design described in reports and detailed drawings. The group that created the design never built a house exactly according to their design. They built many experimental houses, but none of precisely Lo-Cal type.

Actually, the group created a family of four designs, all using similar features and stressing superinsulation. Called A, B, C, and D, these designs are suited to different sizes and orientations of lots. Type A, a single-story house, is the basic design. I describe it first (in the following chapter). The next-most-important design is Type D, a two-story house. I describe it only briefly.

Many other groups (developers, architects, etc.) in USA and Canada have built houses that conform closely to the Lo-Cal design. It has been estimated that, since the detailed plans were published in 1976, 500 to 1500 houses of approximately Lo-Cal type have been built or were under construction by 8/31/80. For example, the Hart Development Corp. of Springfield, Virginia, had completed about 12 such houses, and an additional 32 were under construction or planned for early construction. Charles Wallace of Grafton, Illinois, had completed seven and was building several others. Michael T. McCulley of Champaign, Ill., had completed several. Many other examples could be given.

Comparison of Lo-Cal House and HUD-MPS-1974 House

The Lo-Cal House is far superior thermally to the HUD-MPS (Minimum Property Standards) 1974 house, for 4500-to-8000-degree-day locations. Relative to the HUD-MPS-1974 house, the Lo-Cal House reduces the amount of auxiliary heat required by about 70%. The reduction is attributable to many design improvements the most important of which are:

Much better insulation (superinsulation), and

Better distribution of windows (large area on south, small area on north, no windows on east or west)

Origin of the Lo-Cal Design

The design was developed by a faculty team of architects and engineers at the University of Illinois. Prominent on the team were Professors Warren S. Harris, Rudard A. Jones, Wayne L. Shick, and Seichi Konzo.

As early as the 1920's and 1930's this group had been exploring the potentialities of gravity convective flow of hot and cold air in houses. Because few, if any, reliable, low-cost blowers were then available, much attention was given to gravity convection. Some of the engineering data were published in various ASHRAE periodicals and books.

The Illinois investigators continued to explore methods of improving house insulation and heating, and, in 1944, with the creation of the Small Homes Council, the work was intensified. In the following years the group built and tested several experimental houses and issued many publications describing performance, recommended procedures, etc. By 1971 Konzo had become convinced of the need to improve the insulation of ceilings and roofs, and by 1973 he, together with Shick, Harris, and Jones, undertook to make further improvements, which included optimization of the distribution and sizes of windows. They designed a superinsulated house ("Illinois House") and tried to persuade ERDA to fund the construction of such a house. (ERDA declined.)

In January of 1976 these men started their first computer-assisted study. For example, they made detailed comparisons of certain well-insulated houses of their own design and, for comparison, a house built according to the newly created HUD standards. One promising design that had been proposed by the Illinois group was found to require only 1/3 as much auxiliary heat as the HUD design required -- even though the competing designs called for houses of identical size, identical orientation, and identical rate of air change. The crucial differences were (1) the far superior insulation and (2) the far better distribution of window area of the Illinois house -- which had normal area of south windows but greatly curtailed areas of east, north, and west windows.

At this time (1976), the name Lo-Cal was adopted by R.A. Jones to replace the earlier, informal name Illinois House.

During this period (mid-1970s) the Illinois group gave advice to designers and builders planning to construct very well insulated buildings. For example, it gave help to the persons planning the Saskatchewan Energy Conserving House, to Eugene H. Leger who was planning a house to be built in East Pepperell, Mass., to Harry Hart of Springfield, Virginia, who was planning to build 39 superinsulated houses, and to Charles Wallace of Grafton, Ill., who was planning to build several superinsulated houses.

The Illinois researchers have constructed no Lo-Cal house. Their activities consisted mainly of studies, computer analyses, and educational and consulting efforts. Their main reports are:

Technical Note 14: Details and Engineering Analysis of the Illinois Lo-Cal House. May 1979. 110 p. See Bibliography entry S-185.

Council Note C2.3: Illinois Lo-Cal House. Spring 1976. 8 p. See Bibliography entry S-184-C2.3.

TV cassette Solar House Design, Dec. 1976. 48 Minutes.

Set of detailed drawings of Lo-Cal House.

Several Other Pioneering Groups

Mid-American Solar Energy Complex (MASEC): In 1979 this Minnesota group promoted a broad family of single-family-house designs employing such a favorable combination of (a) insulation and (b) size and distribution of window areas that the amount of auxiliary heat needed in winter would be very small. The group established a specific criterion, or upper limit, on amount of auxiliary heat needed: The Solar 80 criterion. To conform to the criterion, a house must lose very little heat -- or may lose much heat provided that the solar collection area is very large.

By the end of 1979 ten teams of subcontractors had developed, or were developing, many specific designs for houses that would conform to the Solar 80 criterion. It is expected that, by September 1980, MASEC will have brochures and simplified plans available for free distribution. It is expected that by September 1980 MASEC will have detailed descriptions and drawings available for each of these types of houses, for about $50.

Saskatchewan group: This important group is discussed in the next chapter.

Other important groups (discussed in later chapter): Adirondack Alternative Energy, Enercon Consultants Ltd., Technical University of Denmark, Norwegian Technical University.

R.P. Bentley: In 1975 Richard P. Bentley (address: Box 786-T, Tupper Lake, NY 12986) published a 96-p., $15.75, book "Thermal Efficiency Construction" which describes many key features of superinsulated houses. Emphasis is on construction of very thick walls and large-area air-to-air heat exchangers.

E.H. Leger: The unique contribution of Eugene H. Leger of East Pepperell, Mass., is described in detail in the next chapter.

Note Concerning Alaska: Mr. Alex R. Carlson of the Cooperative Extension Service, University of Alaska, Fairbanks, Alaska, has written many technical notes on problems confronting persons designing houses for extremely cold regions. His pamphlets on special procedures for insulating walls and ceilings, and for avoiding harmful build-up of moisture, are of the greatest interest.

Note concerning terminology of rate-of-air-change: If, each hour, the volume of outdoor air introduced into a house exactly equals the volume of the house, the rate of change of the air in the house is called one air-change per hour. In fact, there is much mixing of fresh air and old air, and the actual volume of old air that leaves the house each hour is somewhat less than one house-volume. At the end of an hour some of the old air remains -- about 1/e (that is, about 1/2.72... or about 37%) if the mixing process is continual and thorough. If the mixing is not continual and thorough, the fraction may be very different from 37%; it may be larger or smaller, depending on the airflow patterns and on how close the air inlet locations are to the air outlet locations. Thus expressions such as "one air-change per hour" or "0.3 air-changes per hour" are somewhat misleading.

PERFORMANCE SUMMARY BY INTERNATIONAL GROUP

In August 1980 an international group led by A.H. Rosenfeld of the Lawrence Berkeley Laboratory, University of California, summarized the thermal performances of several outstanding superinsulated houses in USA, Canada, and Europe. See Bibliography item R-260. One conclusion from this report is that superinsulation is especially cost-effective. Another conclusion is that the minimum acceptable rate of airchange may be higher than was assumed a year ago.

Chapter 3
SUPERINSULATED HOUSES: THREE HISTORIC EXAMPLES

INTRODUCTION

Here I present detailed descriptions of three historic super-insulated houses: the University of Illinois Lo-Cal House, Leger House, and Saskatchewan House.

Much attention is given to insulation, windows, vapor barriers, air-change, and thermal performance generally. Little attention is given to fittings, finishes, beauty, livability, and general architectural merit.

The following chapter describes -- less thoroughly -- many other superinsulated houses.

The rationale of superinsulation has been discussed in the previous chapter. A retrospective, function-by-function analysis of superinsulated houses in general is presented in Chapter 5.

Trivial Amount Of Auxiliary Heat Needed

Notice that none of the three houses described here has a conventional furnace. None requires a substantial amount of auxiliary heat. In a mild winter, with the house well managed and slightly-below-70°F temperature accepted, no auxiliary heat may be needed. In a cold winter, and with the house imperfectly managed and a temperature of 70°F demanded at all times, an amount of auxiliary heat needed equivalent to 100 gallons of oil (costing, in 1980, about $100) may be needed.

In other words, the amount of auxiliary heat needed is so small that it is hard to specify. The figure is "within the noise", i.e., so small as to be hard to define, hard to measure, and, depending on many details of weather, house management, and the occupant's style of living, may be anywhere from zero to, say, 10,000,000 Btu (equivalent to about 100 gallons of oil, costing $100).

LO-CAL HOUSE

<u>Lo-Cal House Type A</u> — 40°N.

<u>Urbana, Illinois</u>
(120 mi. S of Chicago)

<u>Origin, History</u>: Designed and promoted in Urbana, Illinois, in 1976 and subsequent years. No actual house of this type was built by the group that made the design.

<u>Climate</u>: The designers assumed a location such as Madison, Wisc., which, in 1961, was rated at 7550 degree days. However, the design is proposed for a wide range of degree day values: a range from 4500 to 8000 degree days -- or even greater if certain modifications, indicated in Technical Note 14 (Ref. S-185), are made.

<u>Building</u>: One-story, 56 ft. x 28 ft., 1570-sq.-ft., three-bedroom, wood-frame house with attic space (cold) and crawl space (moderately cold). There is an attached 2-car garage. The effective internal thermal mass of house and furniture is 30,000 lb. The house faces exactly south.

The living room and three bedrooms are on the south side and the kitchen-dining room, entrances, utility room, and two bathrooms are on the north side. There are two entrances, each of air-lock (vestibule) type; one vestibule includes a coat closet and the other which is larger, serves as utility room. When the sun heats the living room, hot air from this room circulates freely into the kitchen-dining room via an 8-ft.-wide, floor-to-ceiling opening. (The high effectiveness of a wide, floor-to-ceiling opening in permitting gravity-convective circulation interchange of hot and cold air has been demonstrated by Weber et al; see Bibl. W-120.)

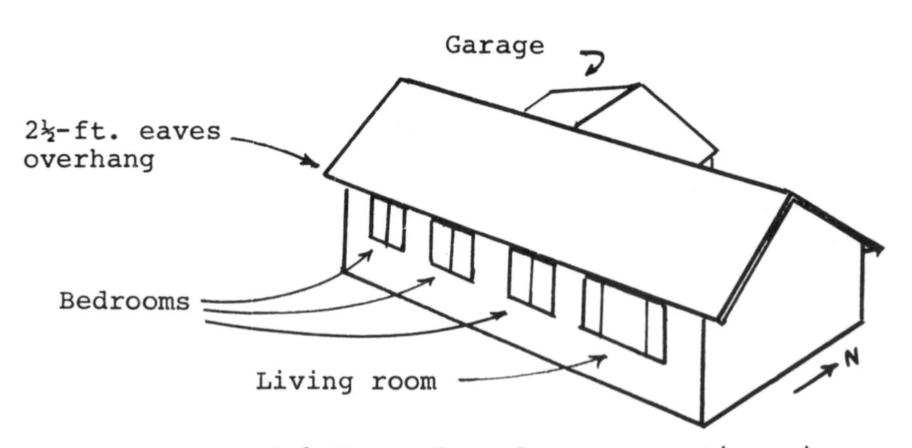

Lo-Cal House Type A, perspective view

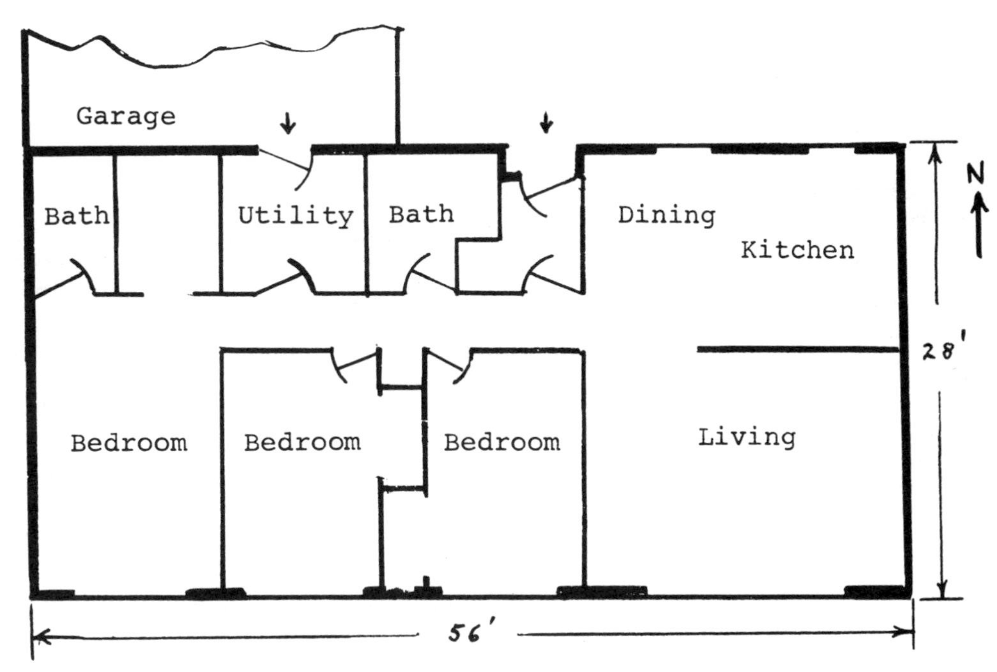

Plan of Lo-Cal Type A House

Windows: The total window area is 144 sq. ft. 85% of this (122 sq. ft.) is on the south side and the remainder (22 sq. ft.) is on the north side. There are no windows on the east or west sides. Note that the area of the south-facing windows is about 8% of the gross floor area. All windows are triple-glazed. The windows are equipped with privacy shades but no thermal shutters or shades.

Walls: Each wall as a whole, i.e., each wall system, includes an outer wall and an inner wall, each employing 2x4 studs 24 inches apart on centers. One set of studs is offset 12 inches relative to the other (that is, shifted 12 inches laterally in direction parallel to the wall) so that there is no all-the-way-through within-stud path for heat-flow. There is a 1½-in. space between the two walls; electric wires run in this space -- with no need to drill holes through the studs. The inner wall spaces, outer wall spaces, and between-wall space are filled with fiberglass batts (batts that have no vapor barrier). The fiberglass is installed in two stages: first it is installed in the outer wall (R-11 or R-13); later, after the installation of electric wires in the 1½-in. space has been completed, R-19 batts are installed in the 1½-in. space and inner wall space (a combined space 5 in. thick).

The vapor barrier, of 0.006-in. polyethylene, is installed on the room side of the inner wall. More exactly: it is separated from the room space by a ½-in. drywall only. If any moisture were to penetrate into the wall system, this moisture could migrate upward and proceed (via the 1½-in. space between inner and outer walls at the ceiling line) into the attic space, whence it would be vented to outdoors. (Note: the blocking installed between the top plates of inner and outer wall is intermittent; thus there is an adequate escape route for moisture.)

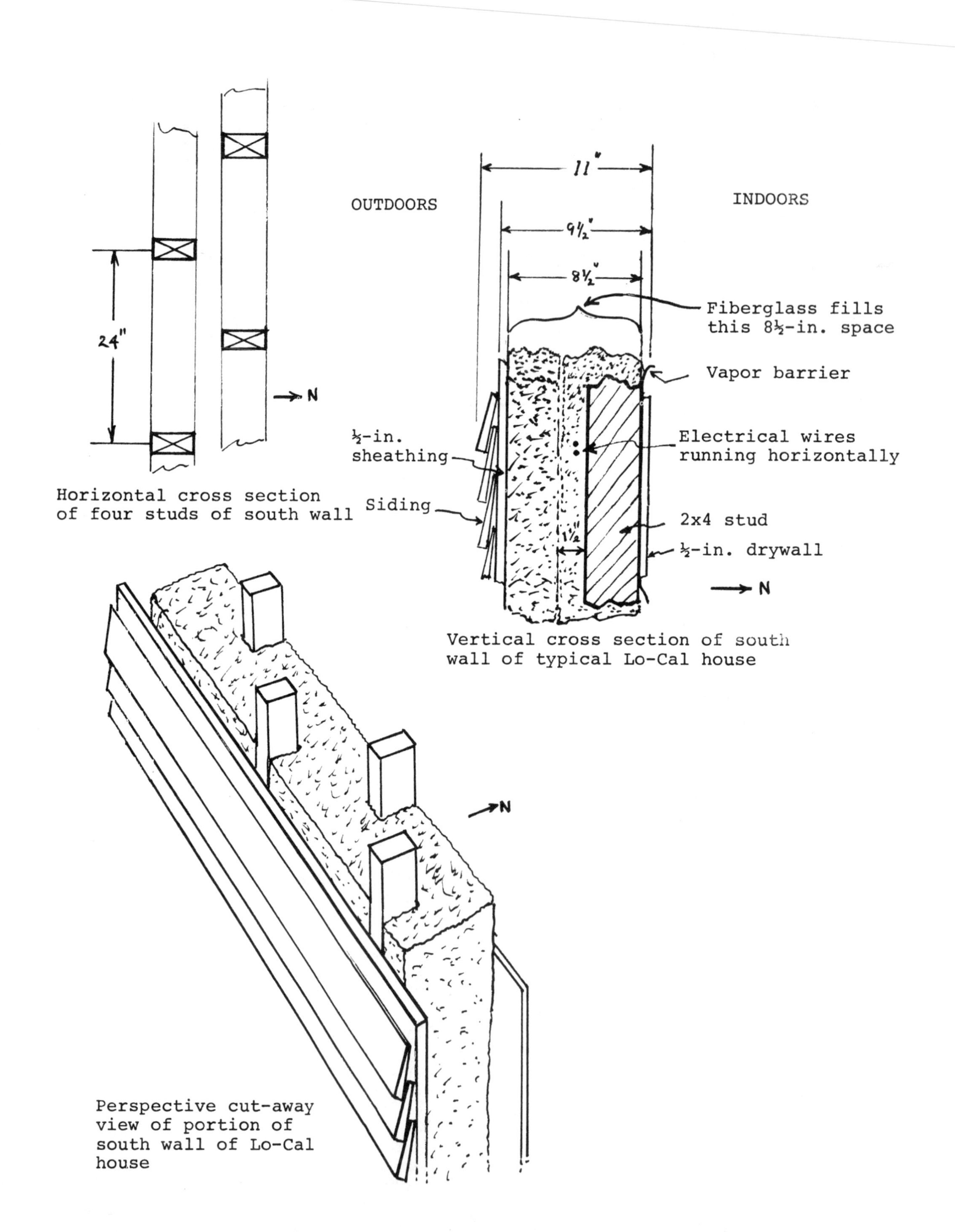

Horizontal cross section of four studs of south wall

Vertical cross section of south wall of typical Lo-Cal house

Perspective cut-away view of portion of south wall of Lo-Cal house

On the outside of the outer wall there is a ½-in. sheathing board, e.g., fiberboard, faced with siding, e.g., wooden clapboards.

The thickness of the wall before drywall or sheathing have been installed is 8½ in. The thickness after these have been installed but before the siding has been installed is 9½ in. The overall thickness, including siding, is about 11 in. The nominal thermal resistance of the complete 11-in.-thick wall system is R-33.

Sills: Fiberglass insulation extends downward to the sills and joins with the top of the sub-floor insulation.

Floor: This includes a vapor barrier and, lower down, a 5½-in. layer of fiberglass. Beneath some areas of the floor there are water pipes; these are situated beneath the vapor barrier but above the fiberglass. The resistance of the fiberglass is R-19. This R-19 insulation also fully covers the band-joist perimeter of the floor system.

Fixed frames of windows and doors: Even here there is considerable insulation. There is a 5½-in. layer of fiberglass except at plates and studs, where there is only 1½ in. A ½-in. plate of Thermax (R-4) is used as a filler between the two 2x6 or 2x8 headers unless the headers are nail-glued plywood headers which have space for fiberglass insulation.

Ceiling: Within the ceiling proper there is a vapor barrier, and above the ceiling proper there is 12 in. of fiberglass (R-38). The ceiling as a whole provides R-39 to R-41.

Eaves: The eaves proper extend 26 in. outward from the outer face of the wall, and the horizontal underside (soffit) is at the same height as the ceiling proper. Accordingly there is some free attic space directly above the wall -- enough space so that the 12-in. layer of fiberglass can extend outward to completely cover the top of the wall system.

Vents: Vents extend along the full lengths of the soffits and the ridge of the roof.

Foundation wall: Each foundation wall is insulated on the inner side with 2 in. of Styrofoam (R-10). The overall resistance of the above-grade portion of the wall is then R-13. Below grade, where the earth helps, the overall resistance is much greater. For the wall as a whole, the effective resistance may be about R-20.

Caulking: Great care has been taken in caulking all remaining cracks.

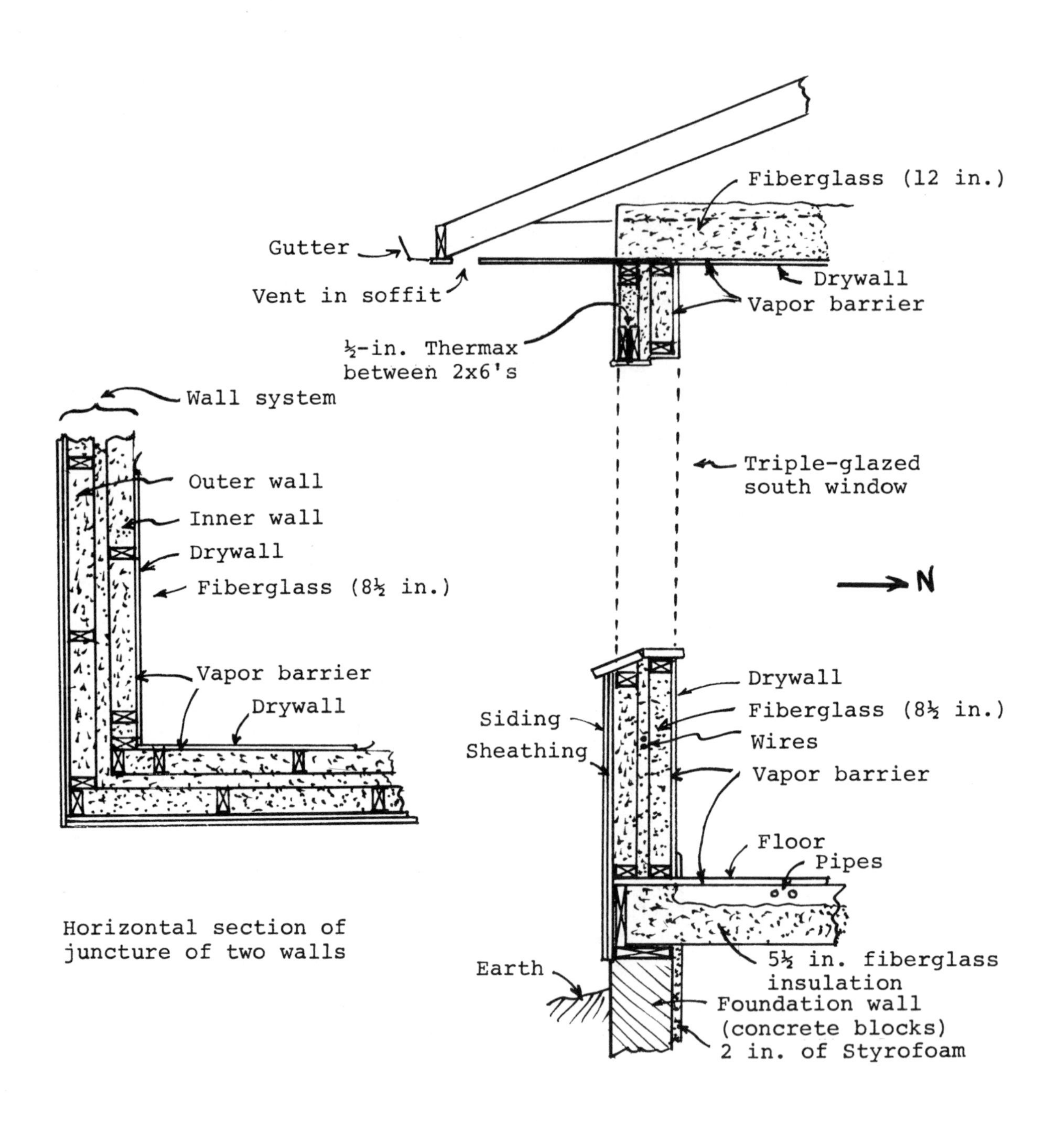

Horizontal section of juncture of two walls

Vertical section of south wall

Lo-Cal House walls: some cross sections

Air change: 0.5 changes per hour is assumed typical. To keep the rate this high may require slight opening of windows or occasional use of forced air change via an air-to-air heat exchanger. If there is no ventilation, the rate may be as low as 0.2 changes per hour -- which under some circumstances may be sufficient.

Humidity: It is believed that no humidity problem exists, inasmuch as (1) exhaust fans in kitchen and bathroom vent much moisture to the outdoors, and (2) there is little tendency for moisture to condense on triple-glazed windows.

Auxiliary heat: Electric.

Domestic hot water heating: Electric, preferably with solar pre-heating.

Solar heating: Much solar energy is collected via the south windows Because they are triple-glazed, they gain more energy during daylight hours of a typical day in winter than they lose during 24 hours. During the heating season, a typical square foot of the triple-glazed south window produces a net gain of about 200 to 400 Btu per typical 24-hr. day, the exact amount depending on the latitude, weather conditions, etc. In regions that have especially sunny and mild winters and have much snow cover, the net gain is much greater.

Shading by the eaves and gutters: The eaves project 26 in. from the outer face of the south wall, and eaves and gutters together project 30 in. The tops of the south windows are 16 in. lower than the eaves, and the window height is 50 in. If the house is at 43°N, these windows are about 90% exposed to direct radiation from Oct. 21 to Feb. 21 and are exposed less than 20% from April 21 to Aug. 21.

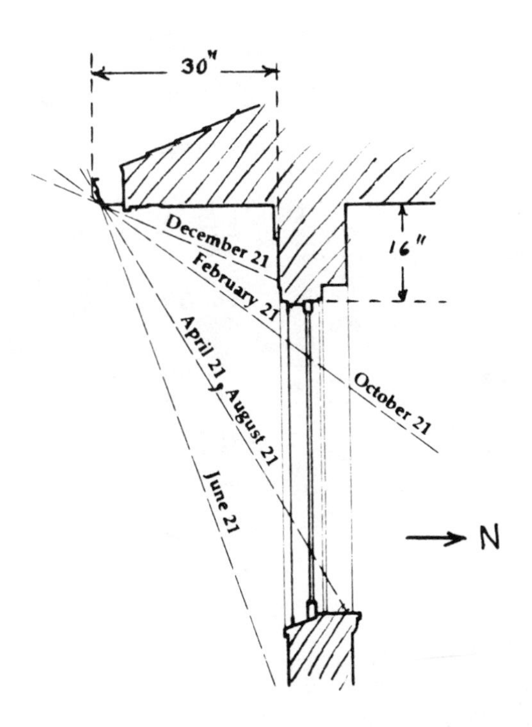

Lo-Cal House: shading of south window by eaves overhang and gutter at noon at various times of year. House assumed to be at 43°N (as at Madison, WI).

Cooling in summer: Any suitable set of air conditioners may be used. A total capacity of 8000 Btu/hr is sufficient at nearly all times. If there are only two or three persons in the house, there may be no need for any air conditioning.

Construction Cost

Believed to be only a few percent greater than that of the HUD 1974 house. The cost of the added insulation etc. is largely offset by the saving from having no furnace and a simpler heat distribution system and the saving from using a smaller air conditioning system.

Performance

Computer calculations have indicated that the greatest amount of heat that would be required from the auxiliary heating system of such a house situated in Madison, Wisconsin, would be 16,500 Btu/hr, or 240,000 Btu/day.

Assuming certain severe conditions (the so-called design heat-loss), one arrives at the following breakdown of heat-losses for this house and for the HUD 1974 house:

	Heat-loss (Btu/hr)	
	Lo-Cal House	HUD 1974 House
Ceiling (1568 sq. ft.)	2980	5960
Walls (1160 sq. ft.)	2670	6960
Floor (1568 sq. ft.)	1570	3140
Windows (144 sq. ft.)	4000	7020
Doors (40 sq. ft.)	200	200
Air change (½ per hour)	8470	8470
	19,890	31,750

For the winter as a whole, these heat-losses and heat-gains have been calculated:

	Loss or Gain (Btu/winter)	
	Lo-Cal House	HUD 1974 House
Loss:	40,400,000	61,900,000
Gain:		
from misc. internal sources	13,100,000	13,100,000
from solar radiation	14,300,000	10,800,000
from auxiliary heater	13,000,000	38,000,000
(Total gain)	(40,400,000)	(61,900,000)

This amount of auxiliary heat for the Lo-Cal House (13,000,000 Btu) would require, for example, burning 130 gal. of fuel oil -- costing, in 1980, in the neighborhood of $130.

The HUD 1974 house would use about 3 times as much oil, and typical houses of comparable size and shape could use 5 times as much, i.e., about $600 worth of oil at $1/gal.

The modest amount of cooling required in summer in the Lo-Cal House would require only about 500 kWh of electricity (to operate conventional air conditioners), assuming that windows are opened on cool nights. If electricity cost 6¢/kWh, the total cost for the summer would be about $30.

Possible Improvements In Design

Many possible improvements are listed in Ref. S-185 (Technical Note 14). For example: adjustable eaves overhang, thermal shutters for windows, reduction of air-change rate to 0.2 changes per hour where circumstances are favorable.

Possible Variations In Design

Permit house to aim as much as 20 degrees from south.

Insulate the walls with cellulose fiber or urea formaldehyde instead of fiberglass. In the ceilings use cellulose fiber instead of fiberglass.

Use Styrofoam or Thermax sheathing instead of fiberboard.

Instead of clapboard siding, use brick veneer.

Replace sheathing and siding with a single sheet of 5/8-in. plywood.

Use wood stove instead of electrical heaters. Or use a coal-burning stove.

Use another kind of heating (e.g., solar heating) for domestic hot water.

Install one or two very small windows on east and west sides of building.

If climate is especially sunny in winter, use larger area of south windows.

Increase the thickness of attic insulation to 18 in. (R-60).

Increase the thickness of the space between the two walls to 3½ in., thus allowing room for an additional 2 in. of insulation, or a total of 10½ in. (R-40).

Provide an air-to-air heat-exchanger.

Lo-Cal House Type D

This house is the next-to-most-important design developed by the Illinois group. It has two stories. Being compact, it is suitable for small lots. Thus it is applicable to townhouses.

The general features are much like those of the Type A house.

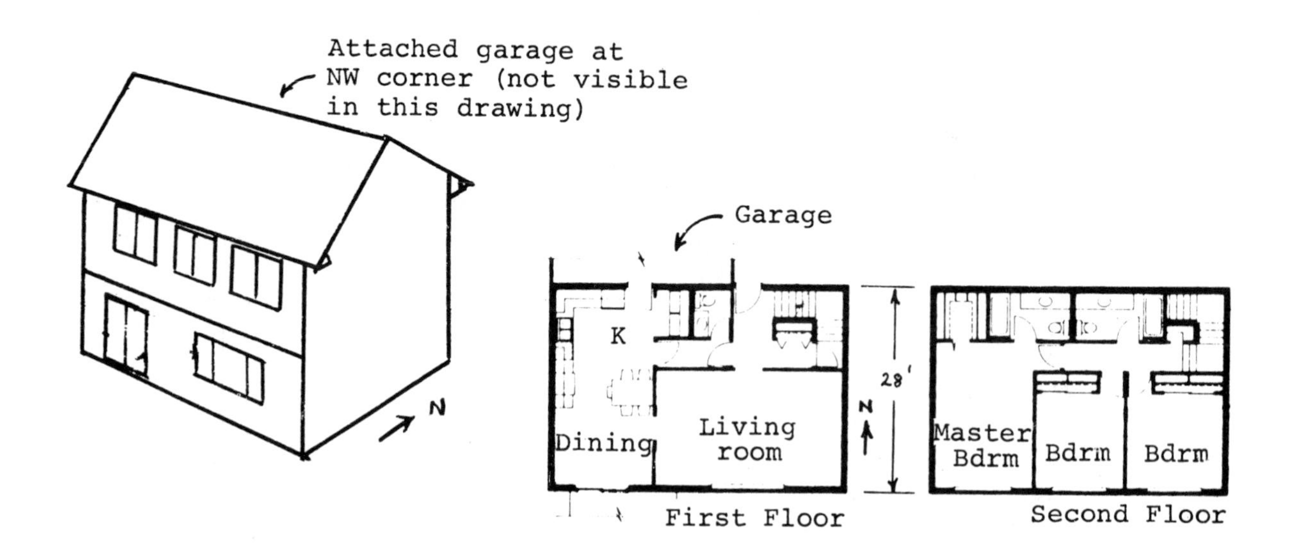

Possible improvements: Install second-story balcony to shade first-story windows in summer. (A balcony is shown in some of the detailed drawings available from the Small Homes Council, University of Illinois.)

See also the list of possible improvements for the Type A house.

LEGER HOUSE
E. Pepperell, Massachusetts
(35 mi. NW of Boston)

See photo. p. 5

Leger House: Hollis St., E. Pepperell, MA 01437.
42½°N. A 7000-DD Location
Occupied late in 1978. Finished late in 1979.

Designer, builder, and original occupant:
Eugene H. Leger, PO Box 95. (617) 433-5702.
Funding: private. Owner and occupant early in 1980: (Name withheld)

Building: One-story, 46 ft. x 26 ft., 1200-sq.-ft., three-bedroom, ranch-type woodframe house with full basement and small attic space. There is no garage. The main story includes a kitchen-dining room, living room, three bedrooms, a bathroom, and two air-lock-type (vestibule) entrances. There is a full basement, reached by a staircase. The (unheated) attic is 5 ft. high at the center and is reached only via a ceiling access panel in the west vestibule. The house faces 20 degrees E of S.

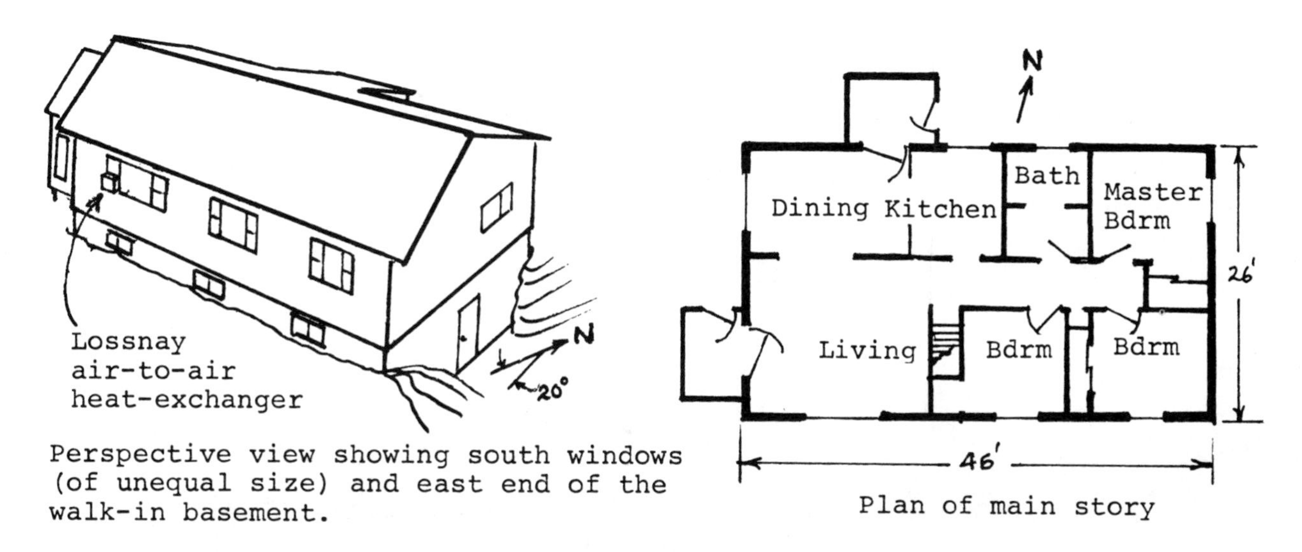

Perspective view showing south windows (of unequal size) and east end of the walk-in basement.

Plan of main story

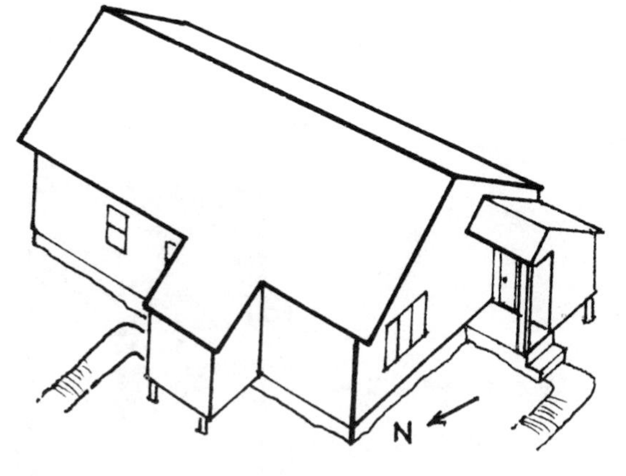

Perspective view showing vestibule-type entrances at SW corner and at north side

Windows: Total area: 153 sq. ft. Of this, 100 sq. ft. is on the south side. All of the windows have frames of wood. About 70% of the window area is triple-glazed, the remainder being double-glazed. There are privacy shades but no thermal shades.

<u>Walls</u>: There is an outer wall and inner wall, each employing 2x4 studs 24 in. apart on centers. One set of studs is offset 12 in. relative to the other so that there is no all-the-way-through within-stud path for heat-flow. There is a 2-in. space between the walls. This space, together with the two sets of spaces between studs, is filled with cellulose fiber. On the room side of the inner wall there is a vapor barrier (of 0.005-in. polyethylene) and two layers of ½-in. gypsum board. On the outside of the outer wall there is a sheathing layer of one-inch. Styrofoam TG (tongue and groove) covered by the siding, which consists of vinyl sheets formed to resemble clapboards.

The overall thickness of the wall is 12 in. Before the siding, sheathing, and gypsum boards were installed, the thickness was 9 in.

The cellulose fiber was installed in the wall system in two stages. The first stage was completed before the vapor barrier and gypsum boards had been installed. The second stage was accomplished after the vapor barrier and one layer of gypsum board had been installed: holes were drilled through the gypsum board and vapor barrier, and the cellulose fiber was blown in. The holes were then closed: they were filled with wooden plugs and were then sealed with taped-on pieces of polyethylene film. The second layer of gypsum board was then installed.

The thermal resistance of the 9 in. of cellulose fiber is about R-36, and the resistance of the complete 12-in.-thick wall is about R-43.

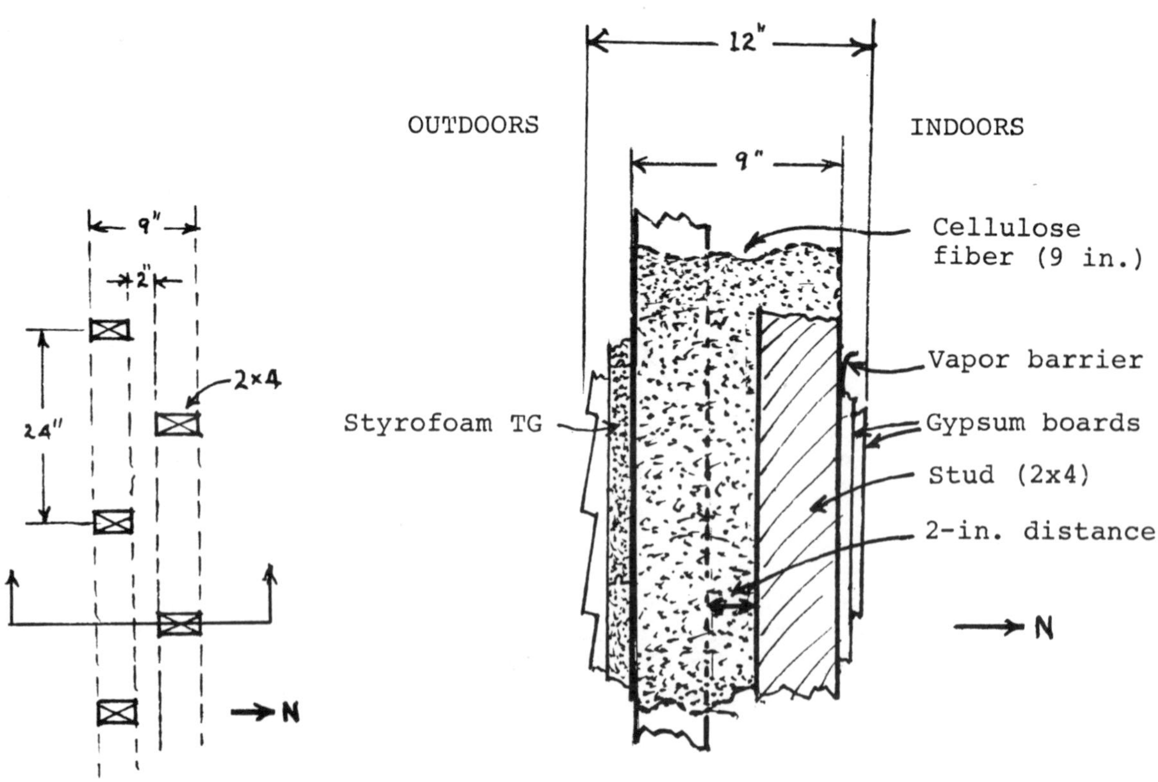

Leger House: Plan view of the two sets of studs, showing offset arrangement

Vertical cross section of a portion of the south wall

There are no holes through the outer or inner walls. None were necessary, because the pertinent electric wires were installed on the room-face of the wall system; specifically, a continuous 3-wire flat vinyl strip (Gould Electrostrip) was applied as a horizontal band to the room side of the inner wall. Electrical outlet boxes were connected to the strip at appropriate locations. The strip was painted in wall-matching color so as to be inconspicuous. (Note: Wires could, of course, be installed conveniently in the space between the inner and outer sets of studs.) Also, there are no pipes in the wall system.

Box beams above windows and doors: Above the windows and the doors there are box beams -- hollow structures (built on-site) that provide much strength but, when filled with cellulose fiber, conduct very little heat. Each box beam, made of 2x4 stringers and stiffeners and ½-in. plywood facing sheets or webs, is about 4½ in. thick. Above each window or door two box beams are used, to provide an overall thickness of 9 in.

Sills: The accompanying diagram shows the construction. Note that there is a moderately large amount of insulation (4 in. of cellulose fiber and 1 in. of Styrofoam TG) outside the joist header. Note also that the Styrofoam covers not only the above-grade walls but also the below-grade walls.

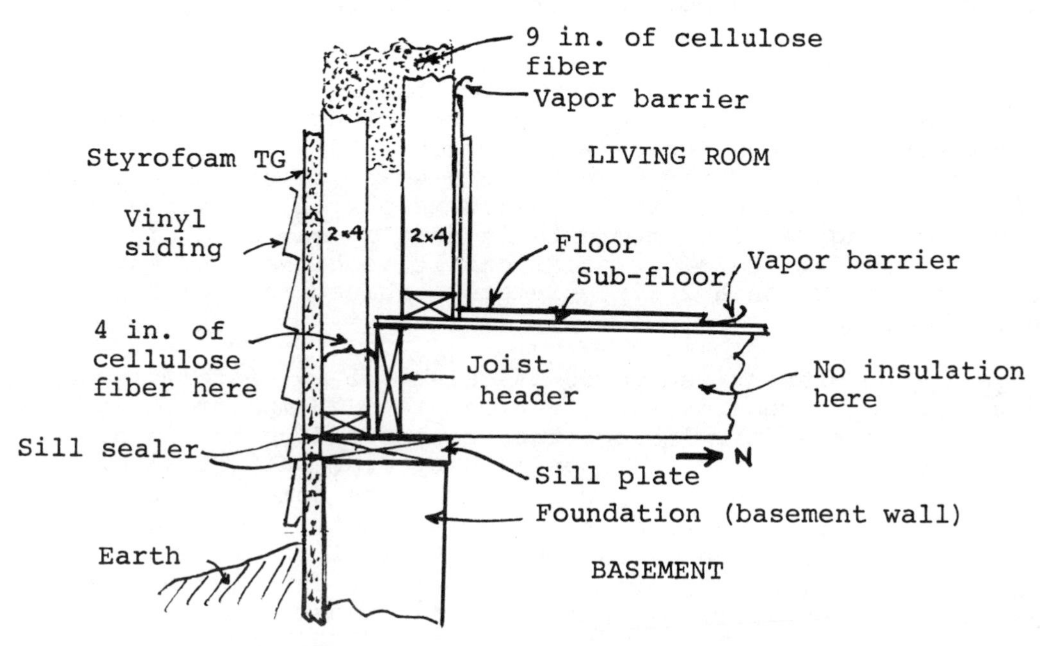

Leger House: Vertical section of sill region.

Roof support: The roof is supported by prefabricated trusses, the load being borne by the external walls. The partition walls carry no load.

Attic: There is 10 in. of blown cellulose fiber resting on the floor of the attic. Just beneath the insulation there is a 0.006-in. polyethylene vapor barrier; no wires pass through it and there is only one hole through it -- a hole for the bathroom vent pipe. This vapor barrier overlaps, and is sealed to, the vapor barrier serving the walls. There are full-length soffit vents and a full-length ridge vent.

Floor: Oak flooring laid on 0.006-in. polyethylene vapor barrier which in turn rests on plywood sub-flooring. Note: The sub-floor was installed before any partitions were erected; thus that floor comprises one large rectangular area with no interruptions. The partitions rest on the finished oak floor, hence could easily be relocated or removed if this were ever desired.

Fixed frames of windows and doors: Precautions were taken to keep heat-loss here small. All cracks were caulked with urethane foam (Poly-Cell-One).

Basement: There is a full basement. It has a walk-out at east end. The basement walls are of 10-in.-thick poured concrete. The S, W, and N basement walls are insulated externally, from top to bottom, with 1 in. of Styrofoam TG. The east wall is partially insulated. The door in that wall is of steel and is insulation-filled. There are three basement windows; all are on the south side, are small, have fixed frames of vinyl, and are double-glazed with Plexiglas.

Vapor barrier: Vapor barriers of 0.005 or 0.006 in. polyethylene are used on all four wall systems of the main story and on the ceiling and floor. All vapor barriers are overlapped and sealed at the overlaps.

Humidity: At most times during the winter the relative humidity remains between 50 and 60%, which makes conditions very comfortable. On a few occasions the humidity reaches 70% but may be reduced by opening windows or turning on a small dehumidifier of conventional type. During most of the summer, with windows and doors open for long periods, humidity is usually satisfactorily low, i.e., below 60%. On a few occasions it is higher and may be reduced by turning on a 20-in.-diameter exhaust fan (with windows or doors open).

Air change: When pressurized to 50 pascals (0.2 in. water), said to be equivalent, as regards infiltration, to a 25-mph or 30-mph wind, the house has an air-leakage rate corresponding to 2.5 air changes per hour, according to tests made by G.S. Dutt et al of Princeton University. (Earlier, when there were some not-yet-sealed cracks, the rate was almost twice as great.) No data are available as to the air-change rate under typical conditions of wind speed and direction; however, it is estimated by the builder that the rate is a small fraction of one change per hour.

Air-to-air heat-exchanger: On 1/2/80 a Mitsubishi Lossnay Model VL-155 MC 56-watt air-to-air heat-exchanger, delivering 65 cfm at 70% efficiency, was installed at the west end of the southwest window.

Domestic hot water: This is heated by a small (knapsack-size) Paloma Constant-Flo gas heater, Model PH-6 rated at 43,800 Btu/hr. It weighs 20 lb. and provides 1.4 gpm of water with 50 F degrees temperature rise. It serves a separate, 40-gal. tank that has 6 inches of insulation. Heat for hot-water taps is provided via a heat exchanger within this tank. The tank supplies hot water directly to the 40-ft.-long baseboard radiator.

Note: There is no conventional chimney serving this gas-type hot-water heater; the combustion products are discharged to the outdoors via a horizontal, 3-in.-diameter, two-foot-long pipe. A small blower in this pipe speeds the discharge; the blower runs whenever the heater runs. The saving from not having a chimney may be about $750.

Auxiliary heat: This is supplied, as a bonus, by the above-described gas heater.

Cooling in summer: A small conventional air conditioning device was installed in July 1979 and, during that summer, was used only for 12 hours in all. On most of the hot days in summer the rooms remained satisfactorily cool (80F or below) if doors and windows were kept shut and shades drawn. A 20-in. electric fan was sometimes operated to cool the house at night. Cooling the house is a slow process.

In periods when the nights also are hot, the house cannot be kept cool without use of an air conditioner. On a few occasions the fan was used during the day to maintain a cooling draft in the rooms. In general the house performs well in the summer and the superinsulation is very helpful.

Construction cost: About $25,000, not including much labor by the designer-builder. The cost is said to be only slightly higher (say, $500 higher) than that of a conventional house of similar size. Although the many special heat-saving features of this house added about $2000 to the cost, there were compensations (especially the saving from having no furnace and no chimney, and from using a much smaller heat-distribution system) of about $1500. The true net incremental cost is impossible to compute accurately; it may be anywhere from $0 to $1000.

Operating cost: During the first 4 months of 1979 -- before certain large cracks had been sealed and before the attic was fully insulated -- the total amount of auxiliary heat used was 11,000,000 Btu, corresponding to burning $40 worth of gas. Other performance figures are presented in Chapter 6.

Performance studies: By G.S. Dutt of the Center for Energy and Environmental Studies, Princeton University.

Proposed improvements:

Use adjustable eaves overhang so that little solar radiation will enter the south windows in the autumn yet much may be allowed to enter in the early spring.

Reduce the south window area about 20%.

Reduce the length of baseboard radiator 70%.

References: Extensive personal communications from the designer-builder. See also "An Affordable Solar House" by E.H. Leger and G.S. Dutt, a paper presented at the October 1979 Passive Solar Conference at Kansas City; see *Proceedings* p. 317. See also an article appearing in the *Solar Utilization News*, Feb. 1980, p. 6, and an article in *New Shelter* for May/June 1980, p. 47.

SASKATCHEWAN CONSERVATION HOUSE

Regina, Saskatchewan, Canada (1000 mi. NW of Chicago)

Saskatchewan Conservation House: 211 Rink Avenue, Walsh Acres (in northwestern Regina) 50½°N. A 10,800 DD location.

Start of operation: Dec. '77. Start of Monitoring: Jan. '78.

Planning and funding:

Research and project management: Saskatchewan Research Council. David Eyre was project manager.

Architect: Hendrik Grolle.

Research assistance:

National Research Council, Div. of Building Research. Harold Orr et al.
University of Saskatchewan, Dept. of Engineering. R.W. Besant et al.
University of Regina, Faculty of Engineering.
Saskatchewan Research Council. David Eyre, Dave Jennings.
Saskatchewan Power Corp.

Development and financing:

Saskatchewan Dept. of Mineral Resources, Office of Energy Conservation.
Saskatchewan Housing Corp. Ken Scherle et al.

Funding:

Province of Saskatchewan through the Saskatchewan Housing Corp.
Department of Mineral Resources, Office of Energy Conservation.

Construction:

Key-West Homes, of Regina, Sask.

Operation and monitoring:

University of Saskatchewan, Dept. of Mechanical Engineering. R.W. Besant, R.S. Dumont, G.J. Schoenau.
Saskatchewan Research Council. David Jennings, David Eyre.
Office of Energy Conservation. M. Edmundson, Fred Hall.

Cost of building:

About $60,000, not including the land or the $15,000 active-type solar heating system or the extensive R&D costs.

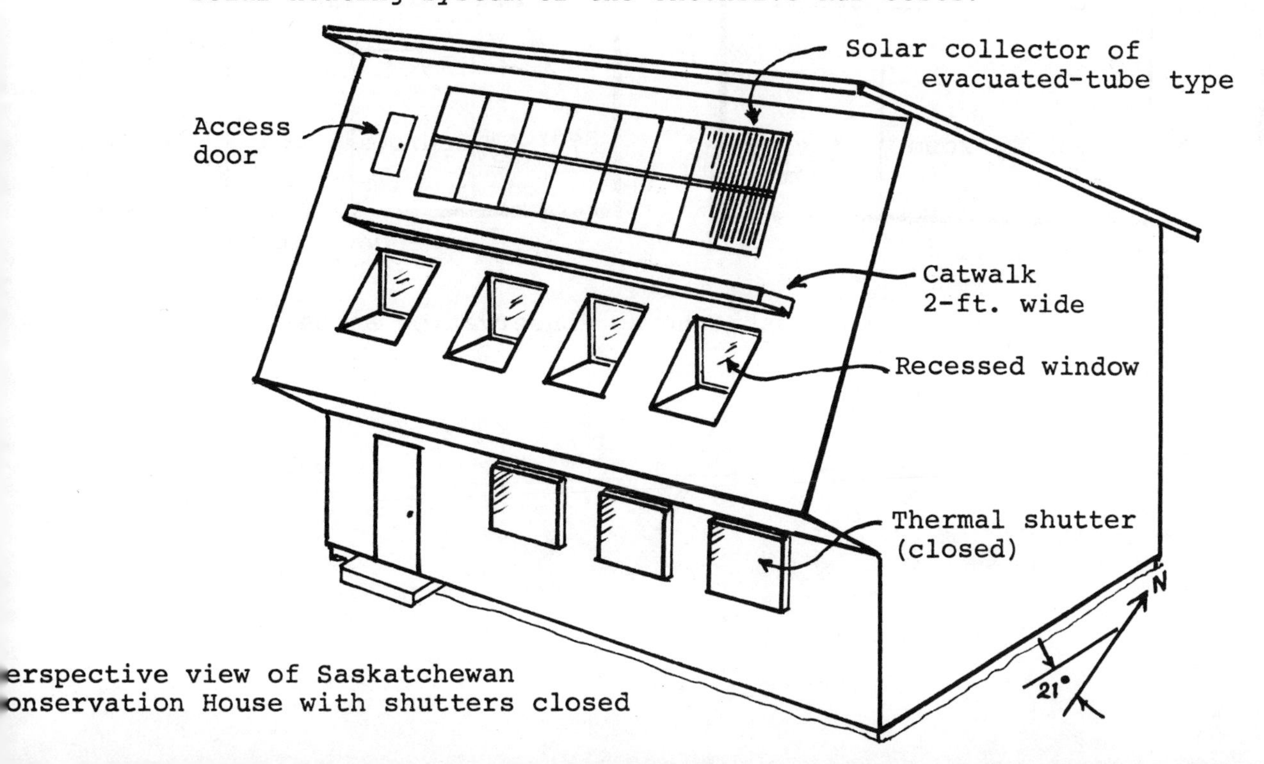

erspective view of Saskatchewan
onservation House with shutters closed

<u>Building</u>: The two-story, 3-bedroom, woodframe house has outside dimensions 44 ft. x 26 ft. The inside dimensions are 42 ft. x 24 ft., with floor area of 2016 sq. ft. There is a crawl space, but no basement and no garage. There is a passive solar heating system and also an active water-type solar system used mainly to serve the domestic hot water system but capable of supplying space heating also. The house aims 21 deg. W of S.

The first story includes family room, living room, kitchen-dining room, bathroom, and two air-lock (vestibule-type) entrances near SW corner and NE corner.

The second story includes three bedrooms, bathroom, den, and laundry.

In the NW corner of both stories there is a thermal storage tank: a two-story-high steel tank for storing hot water.

On the south face of the attic there is a solar collector.

The main roof area slopes 20° upward toward the south, and the south face of the attic is perpendicular to it, i.e., slopes 70° upward to the north. The eaves extend 4 ft. south from the peak of the roof, and a fin-like catwalk just above the second story windows extends 2 ft.

No attempt was made to incorporate large thermal mass in the walls or floors.

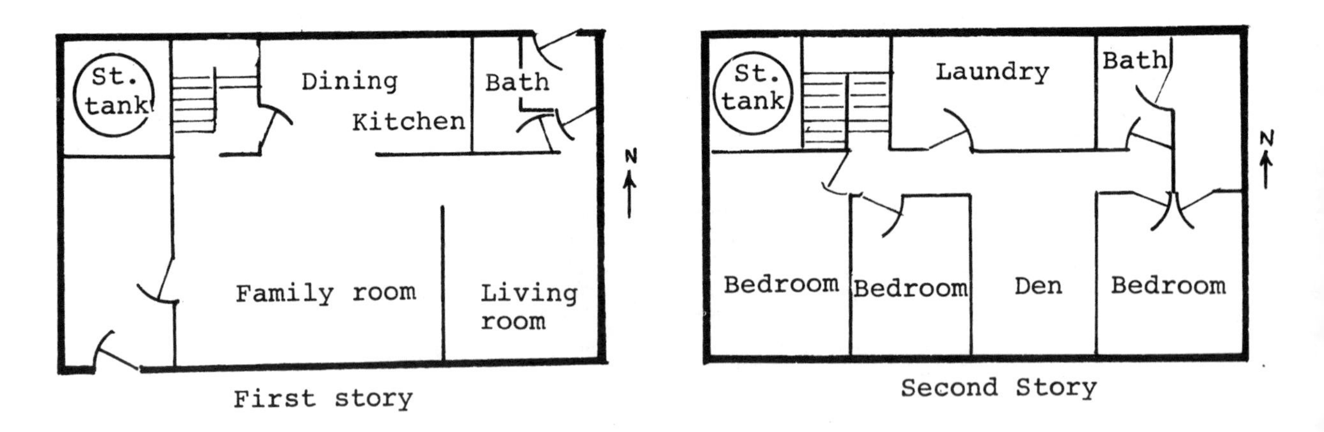

Plans, Saskatchewan Conservation House

Attic: The attic floor is covered with 16 in. of cellulose fiber (R-60).

First story floor: 9½ in. of cellulose fiber (R-36) has been installed between the 2x10 floor joists.

External doors: These are steel clad and contain 1 3/4 in. of urethane. At the edges there are wooden thermal breaks. Air-lock vestibules are used.

Building support posts and sheathing: The building rests on 19 concrete pilings and on 8-in.-thick concrete grade beams insulated on the side toward the building interior with 6 in. of extruded polystyrene foam (R-30). On the outside, 12 in. below the surface of the ground, there is a horizontal apron of extruded polystyrene foam.

Vapor barrier: In the external walls, foundation walls, first-story floor, and second-story ceiling, vapor barriers of 0.006-in. polyethylene have been installed. In each case the sheet is situated on the warm (inner) side of the insulation. Great care was taken to maintain the integrity of the barriers; overlaps and seals were used; perforations were kept to a minimum and were sealed. Molded bags of polyethylene were installed at each electrical outlet box and were sealed to the vapor barrier. All under close supervision!

Air-change: Using a tracer gas technique, with $0^{o}F$ outdoor temperature and with no actively induced air-change, investigators found the change-rate to be 5% per hour. Normally, forced air-change is used, and any desired rate up to 80% per hour may be achieved. Recommended value in mid-winter, with typical family in residence: 20% per hour.

Heat exchanger: In winter much use is made of a 50-to-100-cfm, 80%-efficient, air-to-air heat exchanger, which not only exchanges stale air for fresh outdoor air but also prevents indoor humidity from becoming excessive. For further details as to this heat exchanger, see Chapter 5 and also a 1978 article by Besant et al. (B-251a).

Humidity: If no forced air-change were used, the air-change rate in mid-winter would be only about 5% per hour and accordingly the humidity would soon become much too high. When the heat-exchanger is in routine use, the humidity remains close to optimum.

Passive solar heating: An important amount of solar energy is received (as passive, direct-gain solar heating) via the first- and second-story south windows, which have a total area of 128 sq. ft. The heat thus received is distributed throughout the house by means of a two-speed electric fan.

Windows: The total area is 148 sq. ft. Of this, 128 sq. ft. (87%) is on the south side and 20 sq. ft. (13%) is on the north side. There are no windows on the east or west sides. Half of the window area is double glazed and half is triple glazed.

The area of the south windows is 6.4% of the floor area of the heated rooms.

All windows are covered at night by thick, tightly sealed shutters. The shutters for the first-story south windows are 6 ft. x 4 ft. x 4 in., are Styrofoam-filled, and provide R-20. The edge seals are of rubber tubes. These shutters are hinged at the top and can be swung outward and upward by means of electric motors operated from indoors. The shutters for the second-story windows contain 1 3/4 in. of urethane. They may be slid laterally, manually. When not in use they are stored in slots within the walls. Use of the shutters reduces the overall night time heat-loss of the building by about 30%.

In summer the first-story south windows are shaded by the above-mentioned 4-ft. overhang and the second-story south windows are shaded by the above-mentioned 2-ft.-wide catwalk.

Walls: Each exterior wall has two separate frames. The outer one employs 2x4s, and the inner one employs either 2x4s or 2x6s -- the latter being used on the first story. Within each frame the studs are 24 in. apart on centers. The two sets of studs are aligned -- not offset 12 inches.

Each type of wall includes 13 in. of fiberglass. This is made up of three layers: two 3½-in. batts and a 5½-in. batt. The overall resistance is about R-44. On the warm side of the insulation there is a 0.006-in. polyethylene sheet.

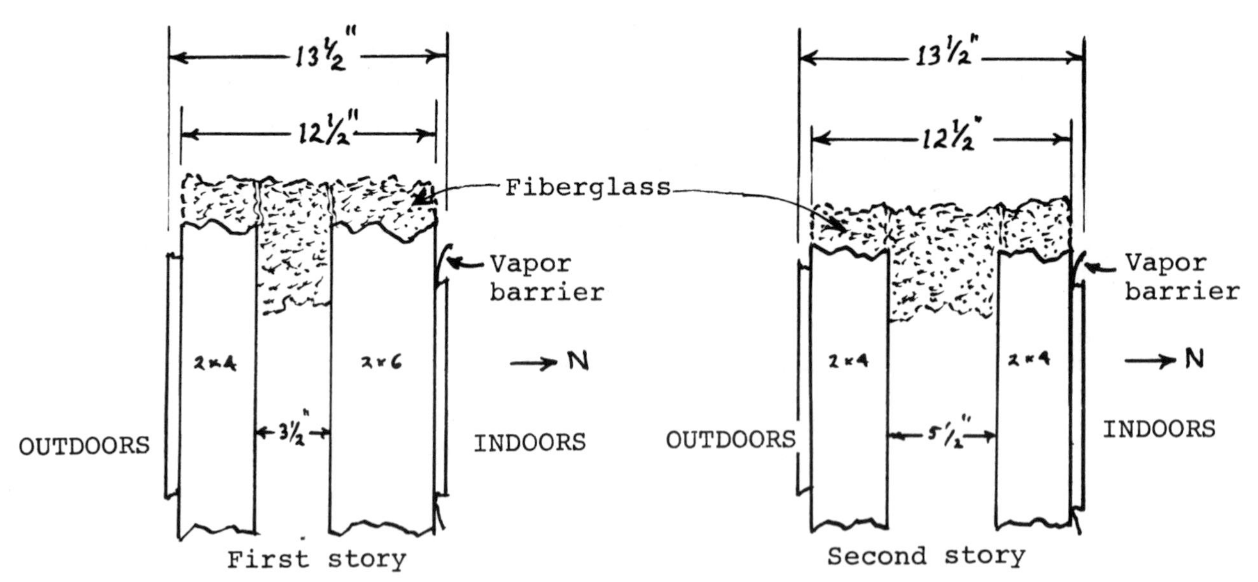

Vertical cross section, looking west, of south wall of the Saskatchewan Conservation House

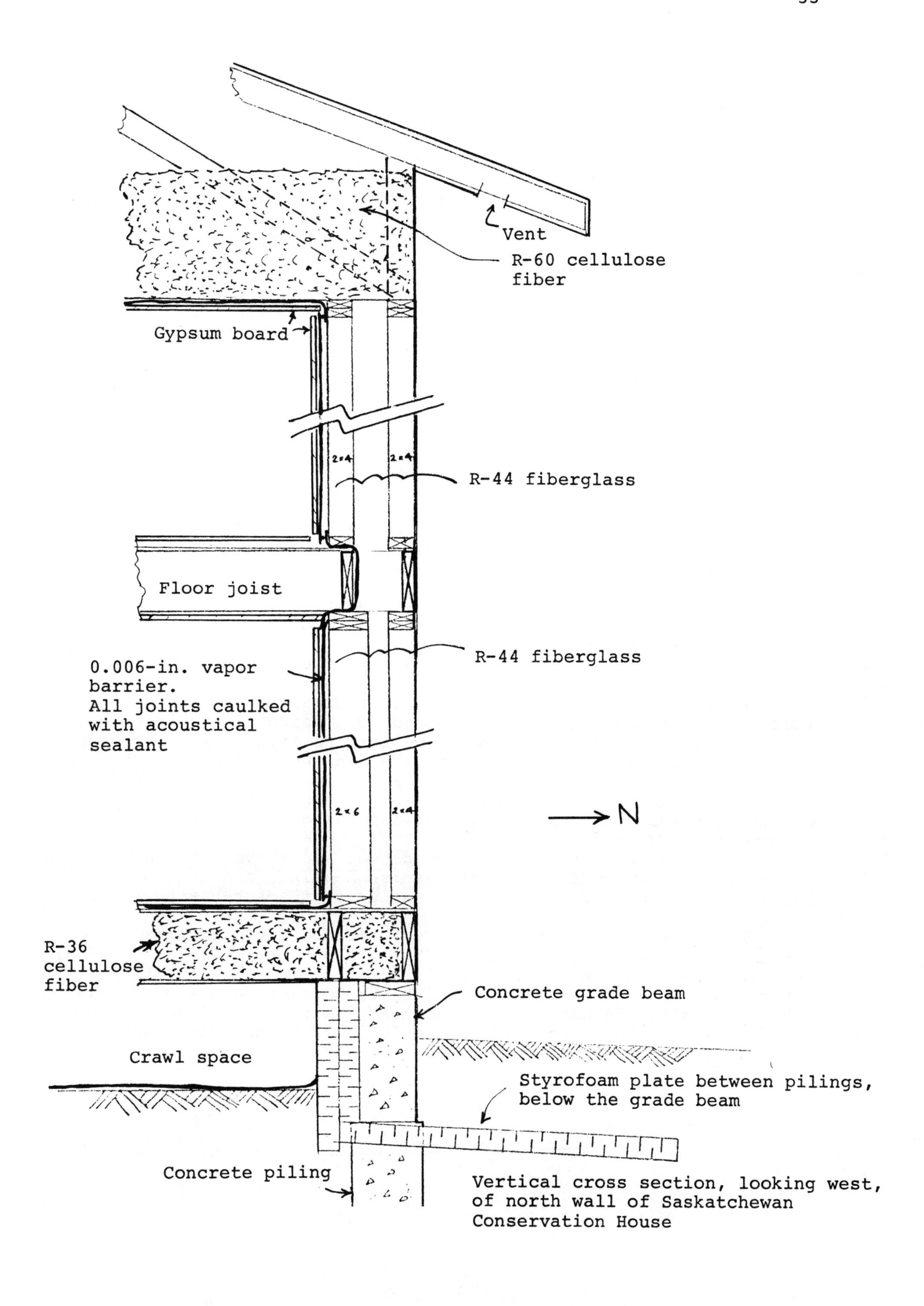

Vertical cross section, looking west, of north wall of Saskatchewan Conservation House

Active Solar Heating System

This is of water type.

Collector: This includes seven 8-ft. x 4-ft. panels of Owens-Illinois water-type collector sloping 70° from the horizontal. Gross area: 256 sq. ft. Net collection area: 192 sq. ft. Each panel includes 12 vacuum-jacketed tubes. The coolant, which is water, is drained at the end of each sunny day. (An earlier arrangement was to use a 67-33 mixture of water and ethylene glycol. Problems arose, and this arrangement was abandoned.)

Storage system: 3360 US gallons of water is stored in a tall (14 ft.) vertical cylindrical 6.5-ft.-diameter steel tank, insulated to R-60, in the NW corner of the house. If initially at 190°F, the tank contains enough heat to keep the house warm throughout a ten-day sunless mid-winter period. The tank includes an electrical heating element, but this has seldom been used. Indeed, the storage system itself has been put to little use other than heating the domestic hot water.

Auxiliary heat: There is no oil, gas, or wood back-up heating. When and if passive solar heating combined with intrinsic heat is not sufficient, heat may be taken (by a fan-and-coil) from the main thermal storage tank. Alternatively, one may merely turn on the top burners of the electric stove: they are adequate even for offsetting the peak heat-loss of the building, which, at -30°F, is 3.7 kw, or 12,600 Btu/hr.

Domestic hot water: A 65-gal. stone-lined tank with internal heat exchanger and back-up electrical heating elements is used. Heat is transferred to this tank from the main, solar-energy-heated, storage tank by a centrifugal pump controlled by a thermostat.

Performance: The significance of the performance data has been reduced by the fact that many thousands of visitors have inspected the building and have opened doors numberless times. In other words the conditions of use have been abnormal in the extreme.

Nevertheless, in the period since monitoring was started (Jan. 1978), R.W. Besant et al of the Mechanical Engineering Department of the University of Saskatchewan have amassed much interesting information.

The annual amount of heat needed over and above what is supplied by intrinsic heat sources, passive solar heating, and active solar heating is: none. In other words the percent-solar-heated figure is 100%.

The annual amount of heat needed over and above intrinsic heating and passive solar heating (but without active solar heating) is about 3700 kWh (about 12.3 MBtu). (Note: The investigators found, in the winter of 1978-79, with the house unoccupied and virtually no intrinsic heating, the heat-need was about 10,000 kWh or about 34 MBtu. Had there been a family in residence, the heat-need would have been only about 1/3 as great.)

On a typical night in winter, the actual gross rate of heat-loss was 133 Btu/hr per F-degree of difference between indoor and outdoor temperature; i.e., 70 W/°C.

The calculated value of effective thermal capacity of the house (not counting the large storage tank) is 10,800 Btu/°F; i.e., 20.5 MJ/°C.

The time-rate-of-cooldown on a cold (-10°F) night with no intrinsic heating, with the shutters closed, and the thermal storage tank not in use, was found to be 0.8°F per hour. This figure is consistent, within about 20%, with the measured value of heat-loss rate and the calculated value of effective thermal storage mass of the house proper.

During a night in which there is no intrinsic heating, the shutters are closed, and the thermal storage tank is not in use, the 1/e time constant of cooldown is about 100 hours. (Note: the individual time constant of the ½-in. gypsum boards is about ½ hour, and nearly all pertinent components of the house have individual time constants of less than 2 hours. Thus the cooldown of the house as a whole, on a single night and with no intrinsic heating and no use of the thermal storage tank, closely follows a simple exponential curve.)

Recommended <u>changes</u>: In the light of trial use of this house, and related studies, these changes are currently recommended, by the engineers monitoring the performance, to persons considering building houses of this general type:

Omit big active solar heating system. Although designed primarily for space heating, it is not significantly needed for this purpose. Its main use is in heating domestic hot water -- which could be accomplished by a much more modest solar-heating system.

Use a different location for the vapor barriers serving the walls. Place them on the <u>outer</u> face of the inner set of studs -- to avoid interference with the electrical wires. See U-915.

<u>References</u>: Many excellent reports are available:

B-250 Besant et al: "The Saskatchewan Conservation House: Some Preliminary Results".

B-251 Besant et al: "The Passive Performance of the Saskatchewan Conservation House".

B-251a Besant et al: "An Air to Air Heat Exchanger for Residences".

U-915 University of Saskatchewan: "Low-Energy Passive Solar Housing Handbook".

B-251c Besant et al: "The Saskatchewan Conservation House: A Year of Performance Data".

For further details concerning these reports, see the Bibliography.

I am indebted to R.S. Dumont for supplying me with much additional information, including information up-to-date as of January 1980.

Chapter 4

SUPERINSULATED HOUSES: OTHER EXAMPLES

ALASKA

Fairbanks (In center of Alaska)

Roggasch House: 215 Ina St., Fairbanks, AK. 65°N. A 14,200 degree-day location. Completed early 1980.

Designer, builder: Bob Roggasch. House built for sale.

Warning: My information concerning this house is very limited.

Building: One-story plus basement. Wood-frame house. 2000 sq. ft. Has 16 in. of cellulose fiber. Beneath the 4-in. concrete slab of the basement there is a 6-in. layer of gravel and 6 to 7 in. of urethane foam. Window areas on all four sides are very small. Eaves overhang: about 1 ft.

Performance: The house is expected to have a heat-loss of less than 6000 Btu per sq. ft. per winter. The predicted amount of oil needed for auxiliary heat (from oil-fired water heater in basement) per winter is 220 gal.

Cost: Selling price of the house may be near $74,000. The added insulation accounts for about $6000 of this.

Reference: Fairbanks Daily News-Miner, Jan. 19, 1980.

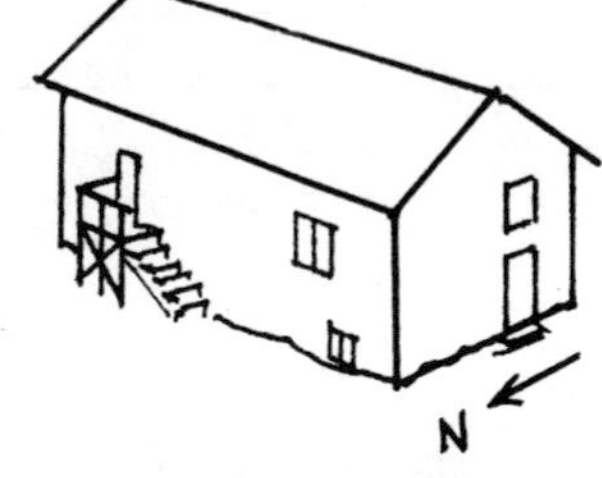

ILLINOIS

Champaign, Illinois (120 mi. south of Chicago)

McCulley House: 4003 Farmington Dr., Champaign, IL., Jan. 1979. 40°N., 6000 degree-day location.

Designer, builder, owner, occupant: Michael T. McCulley.

Building: One-story, 68 ft. x 28 ft., 2100-sq.-ft., 3-bedroom, 2-bathroom, wood-frame house with air-lock (vestibule) entries, attached two-car garage, and no basement. There is a crawl space. The house is 6 ft. extra wide at east end to allow space for family room. Windows (206 sq. ft. on south, 45 on north, none on east or west) are triple-glazed. Insulation: south wall, R-14, north, east, and west walls, R-33. Ceiling: R-44. Floor: R-19. A 0.006-in. polyethylene vapor barrier is used on ceilings, walls, and crawl-space floor. There have been no moisture problems to date, especially inasmuch as fresh air is drawn into the house when bathroom vents are running or furnace is on. The house has soffit vents and also clerestory vents. Auxiliary heat provided by gas furnace and fireplace. Central air conditioning. Cost: About $\$35/ft^2$.

The design generally conforms to the University of Illinois Lo-Cal House.

Typical cost of fuel (gas) for auxiliary heat on mid-winter day: $1/day (in 1979, per article in the Apr. 1, 1979, issue of Champaign-Urbana News-Gazette.

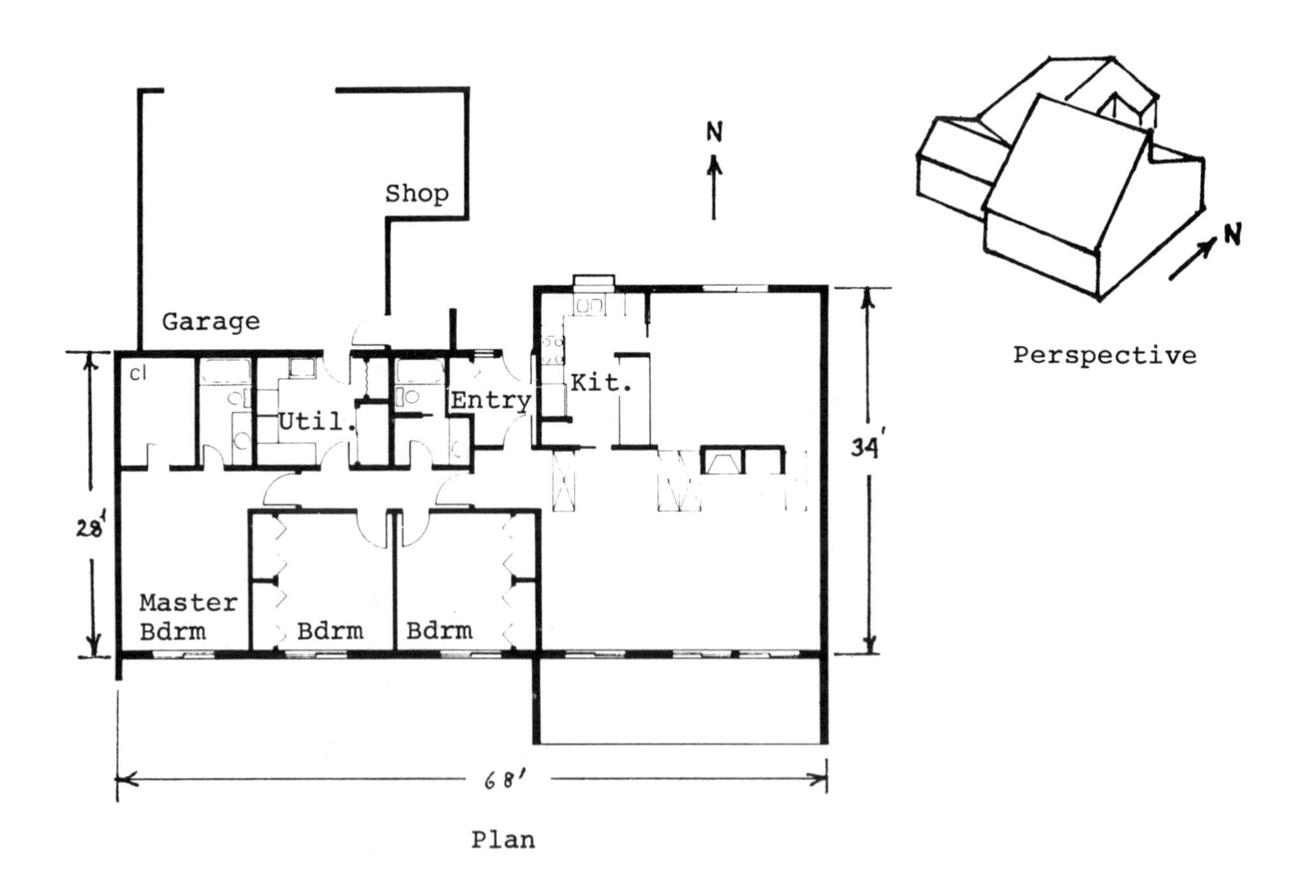

Plan

Perspective

Champaign, Illinois (120 mi. south of Chicago)

Laz House: 4001 Farmington Dr., Champaign, IL., Oct. 1978. 40°N., 6000 degree-day location.

Designer, builder: Michael T. McCulley

Building: One-story, 66 ft. x 28 ft., 1850 sq. ft., 3-bedroom, 2-bathroom, wood-frame house with air-lock (vestibule) entries, attached, two-car garage, and no basement. Windows (210 sq. ft. on south, 63 on north, 21 on east, 21 on west) are triple-glazed. Insulation: south wall, R-20, north, east, west R-28. Ceiling: R-44. No insulation in floor. Foundation wall: R-20. Auxiliary gas-hot-air system; also fireplace. Design is approximately of Lo-Cal type. Many other design features are same as for McCulley House.

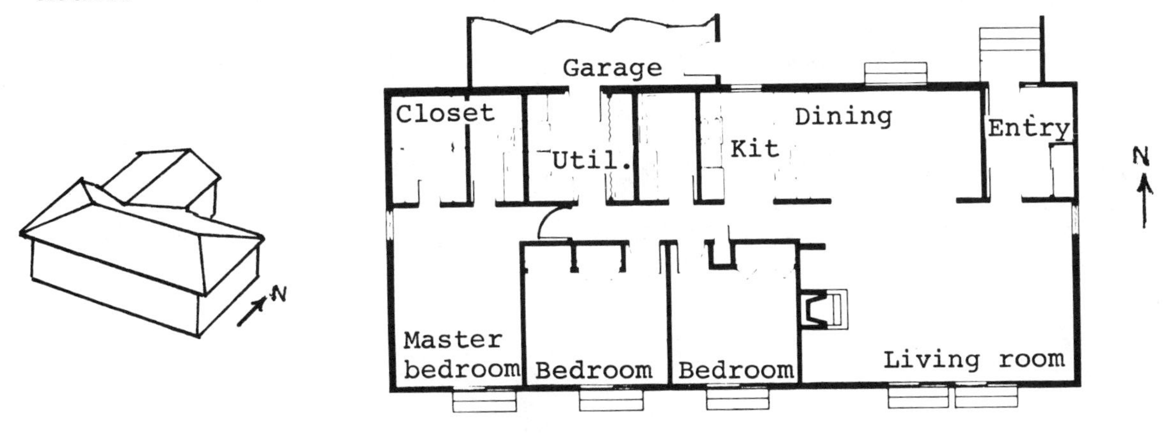

Urbana, Illinois

Duplex Building: 507 East Scoville St., Urbana, Ill. 1977. 40°N. 6000 DD.

Designer, builder: Michael T. McCulley

Building: Includes two dwellings: east and west. Each is a one-story, 40 ft. x 28 ft., 1200 sq. ft., wood-frame construction, with 2 bedrooms, 1 bathroom, and an attached 1-car garage. Crawl space; no basement. Window area: south, 90 sq. ft., north, 24, outer end wall, 12, other end wall no window (party wall). Insulation: wall, R-33; ceiling, R-44; floor, R-19; foundation wall, R-8. Auxiliary heat: electric. Central air conditioning. Cost of auxiliary heat in winter of 1978-79: about $70. The design conforms closely to the Lo-Cal design.

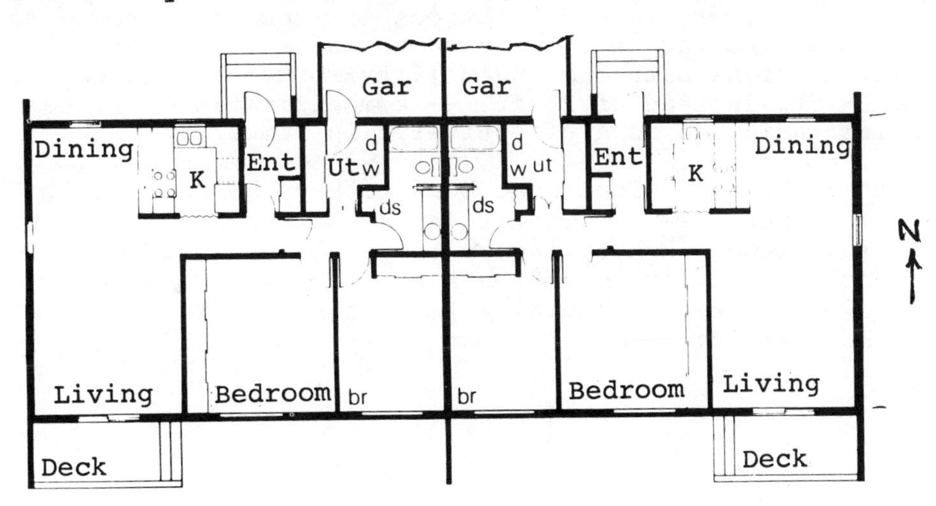

Lerna, Illinois

Phelps House: RR #1, Lerna, IL 62440. Nearly complete Nov. 1979.

Owner, occupant: Richard A. Phelps. Note: most of the labor was supplied by owner and friends.

Building: A one-story, woodframe, superinsulated house based on Lo-Cal design. Ceiling, wall, and floor are insulated to R-60, R-40, and R-32 respectively. Walls of perimeter wall serving crawl space are insulated. Extensive use of polyethylene vapor barriers. Window area is small: 120 sq. ft. in all. (45 on S, 55 on E, 20 on W, 0 on N). All windows Andersen triple glazed. Solar energy plays a minor role. Auxiliary heat: electric.

Performance: Auxiliary heat requirement: about 1.1 Btu/(sq. ft. DD). Almost no cooling needed in summer. (11,000 Btu/hr air conditioner is on hand.) Comment by W.L. Shick: "The house is an outstanding example of super-R at very low cost, using standard platform framing, studs, window headers, trusses, electric wiring, wall height, etc. -- no extra layers of sheathing or drywall." The house might have performed even better "..with more south windows and less east windows." Total electrical power used specifically for room heating: about 3000 kWh per winter, costing about $120 at 4¢/kWh. No air-to-air heat-exchanger was used, yet no problems of humidity of odor were encountered.

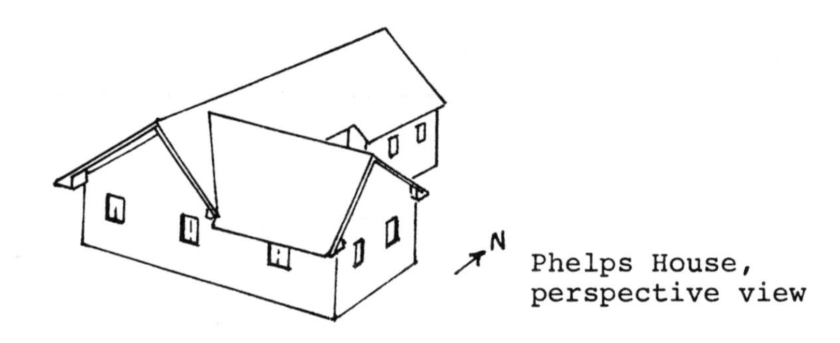

Phelps House, perspective view

Macomb, Illinois

Weller House: 1407 Joseph St., Macomb, IL 61455. Dec. 1977.

Owner: Richard D. Weller.
Cost (incl. land but not incl. labor): About $50,000.
Estimated value of labor: $7000 to $12,000.

Building: This is a modification of a Lo-Cal Type A house, described in detail in Chapter 3. Modifications included (1) providing two large bedrooms instead of three standard-size bedrooms, (2) reversing the positions of NW bathroom and utility room, (3) making the eaves overhang 3½ ft. instead of 2½ ft. -- a modification that tends to exclude solar radiation more effectively (in summer).

Performance: Cost of electrical heat to keep house at 70 to 72F day and night: about $340 per winter (at 4½ cents/kWh). Amount of air conditioning needed to keep indoor temperature at 80F or below in summer: none. (An air conditioner is on hand and has been used on rare occasions to reduce humidity.) In three years, no moisture problems have arisen.

IOWA

Cedar Rapids, Iowa (105 mi. ENE of Des Moines)

Kirkwood Energy Saver 80 House Type 1A: Q Ave. at 27 St. NW. A 6700 DD location. 42°N. Completion: Feb. 1980.

Sponsor and owner: Kirkwood Community College.
Program manager: Larry Bean, Passive Solar House Construction and Education Program.
Thermal design advisor: D.A. Robinson, Mid-American Solar Energy Complex.
Design team coordinator: Ed Sauter of Mount Vernon, Iowa.
Construction team coordinator: Ralph Driscoll, instructor of the residential carpentry program.
Funding of design, education, and evaluation: Mid-American Solar Energy Complex, a regional solar center of DOE. Funding of building proper: about $70,000.
Building may be sold to private individual by summer of 1980.

Building: A one-story, 1200-sq.ft., 50 ft. x 24 ft., three-bedroom, two-bathroom woodframe house with full basement and attached two-car garage. Doors are of vestibule type. General design of house conforms to Univ. of Ill. Lo-Cal scheme and MASEC Solar 80 auxiliary heat-need criterion.

Windows: 110 sq. ft. on south, none on east, about 30 on north, 10 on west. South windows are double-glazed, north windows triple-glazed.

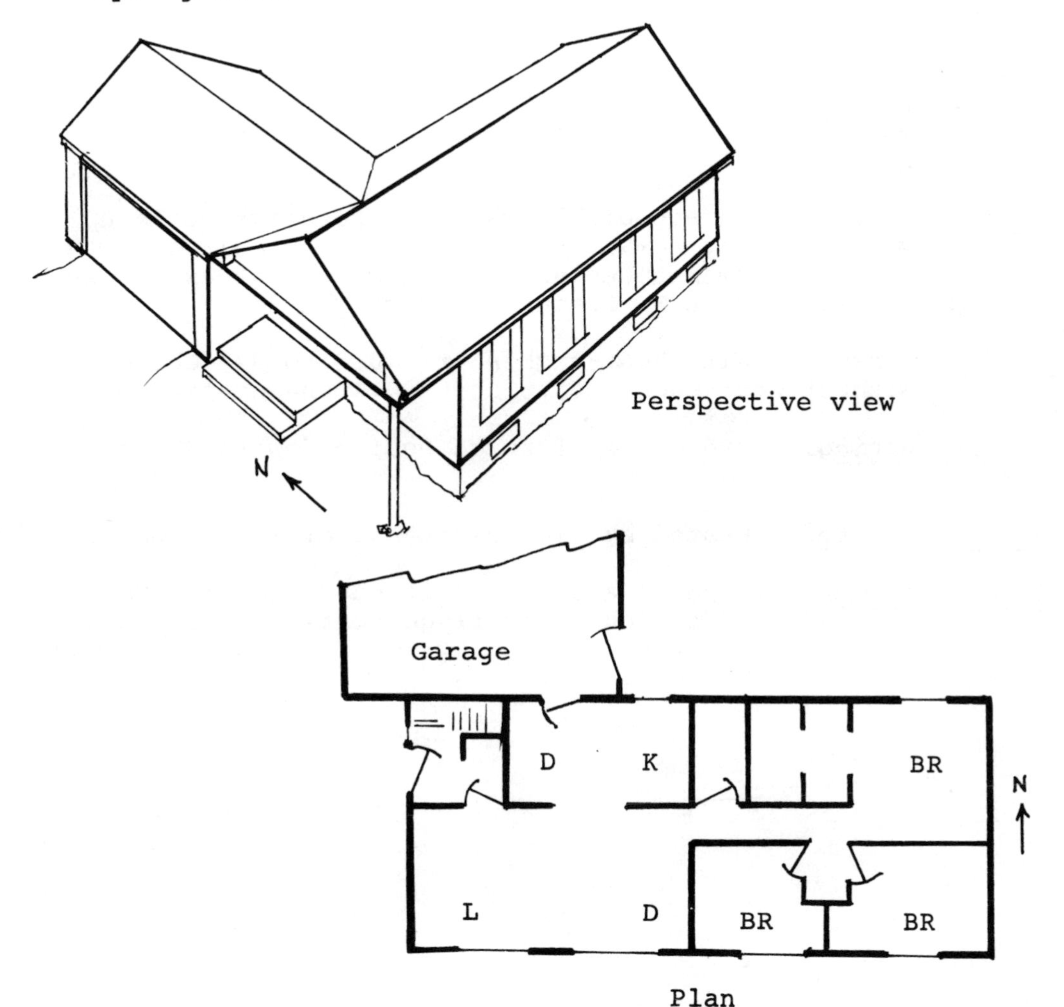

Perspective view

Plan

North wall: Has two sets of 2x4's, with 2-in. separation, to allow room for 3½ in. and 5½ in. fiberglass batts; the total of 9 in. of fiberglass provides R-30. On warm side of wall there is a 0.006-in. polyethylene vapor barrier, protected by 5/8-in. drywall. Overall thickness of wall, including drywall and sheathing: 11 in.

South wall: Employs 2x6 studs, 5½ in. of fiberglass, plus 3/4 in. rigid insulation. Vapor barrier protected by drywall. R-26.

Ceiling: 12 in. of fiberglass (R-38) rests on vapor barrier.

Floor: Not insulated.

Basement: The 8-in. concrete walls are covered on outside with 2 in. of Styrofoam. Above-grade areas, and areas extending down 1 ft., are coated with fiberglass-reinforced surface-bonding cement. Floor is a 4-in. concrete slab resting on vapor barrier on 4 in. of gravel fill.

Eaves: Eaves and gutter extend 2½ ft. outward, to shade windows in summer.

Vapor barriers: 0.006-in. polyethylene -- on walls, ceiling, basement floor.

Vents: Full length soffit vents and ridge vents. Exhaust fans in kitchen are ducted to air-to-air heat exchanger mentioned below.

Thermal mass: Basement walls insulated on exterior, basement floor slab of concrete, first-story partition walls of 4-in. solid concrete blocks, double thickness of sheetrock on east and west walls.

Humidity: Stays within acceptable range. Excessive build-up of H^2O is avoided by air-to-air heat-exchanger associated with the venting of air from kitchen and bathrooms. Heat-exchanger is of type developed by R.W. Besant and Dick Van Ee.

Rate of air change: With heat-exchanger: 0.3 to 0.5 changes per hour. Without heat-exchanger: about 0.1 change per hour.

Auxiliary heating: By electric furnace equipped with 3.5 and 5.0 kW coils.

Domestic hot water: Heated by electricity at off-peak hours.

Cooling in summer: There is a one-ton air conditioner. No attic exhaust fan. Soffit vents and ridge vents provide much natural ventilation of attic.

Cost of passive solar heating system: About $3000 for materials and labor.

Performance: Detailed data not yet available. It is expected that the gross heat-loss per winter will be 16,000,000 Btu and that half of this will be made up by the combination of intrinsic heat sources and direct passive solar gain. The other half will be provided by the electric furnace -- which may provide 2400 kWh (worth $100 at 4.2¢kWh; corresponding to 80 gal. oil). In a 24-hr. test in mid-January, with outdoor temperature about 5°F, with no intrinsic heating of any kind, room temperature fell from 70°F to 62°F at night and, in the subsequent partly cloudy daytime, rose to 66°F -- an overall drop of only 4 F degrees.

References: 14-p. report scheduled for publication in Solar Age, April 1980. Also a 15-p. set of whiteprint plans (24" x 18") describing the house in great detail: drawings and specifications. Also personal communications.

MINNESOTA

Northfield, Minnesota (25 mi. S of Minneapolis)

Saltbox House: 44½° N. An 8000-DD location. Occupied in Nov. 1979.

Designer, builder, owner, occupant: Michael G. Scott. Funding: private.

Building: Two-story, 2200-sq.-ft., 3-bedroom, 1½-bathroom, woodframe, saltbox-type house that is oriented at 45° from south (due to lot lay-out problem). There is an attached 2-car garage at NW corner. The ground floor includes living room with fireplace, dining room, kitchen, dinette, library, ½ bathroom, and, at NW corner, a vestibule-type passage to outdoors and to garage. The second story includes 3-bed-rooms and a bathroom.

Windows: 275 sq. ft. in all. This includes 195 sq. ft. on SW and SE sides, and 80 sq. ft. on NW and NE. Glazing is double on SW, SE, and triple on NW, NE.

Walls: Employ 2x8s 24 in. apart on centers. Include 7½ in. fiberglass. Sheathing is of 1-in. Styrofoam. Frames of doors and windows are sealed with urethane foam from spray kit. All electrical and plumbing lines are well sealed. On warm side of wall there is a 0.006-in. polyethylene vapor barrier. Foun-dation walls are insulated on outside with 2 in. of Styrofoam down to frostline and 1 in. down to footing. Typical R-value of wall: R-30.

Ceiling: Cellulose fiber provides R-60.

Air-to-air heat-exchanger: This is connected so as to scavenge heat from air being vented from bathroom and dryer.

Auxiliary heat: From wood stove and fireplace.

Cost of construction: It is estimated that the superinsulation added 2% to the cost of construction, i.e., added about $1/ft². It is believed that such spending for superinsulation is far more cost-effective than spending for a large solar heating system of passive or active type.

Expected performance: It is predicted that the heat-loss during an hour when the outdoor temperature is -10°F will be 17,000 Btu/hr -- and that the heat-need for the winter will be 1/5 that of a currently-built con-ventionally heated house. On a 32°F night, with no intrinsic heating, the house cools down very slowly: the temperature of room air drops only 6 F degrees in 12 hours. In normal use, the house requires no auxiliary heat during a 24-hr. night-and-day period in which the average outdoor temperature is 32°F or higher. Direct gain passive solar heating provides 30% of the heat needed to compensate for heat loss. If all of the heat needed from an auxiliary system were supplied by electrical heaters, the cost of the electricity per winter would be about $200 (to be compared to $1000 for comparable house not super-insulated).

References: MASEC News, Jan. 1980. Northfield (MN) News, 2/7/80. Also personal communication from owner.

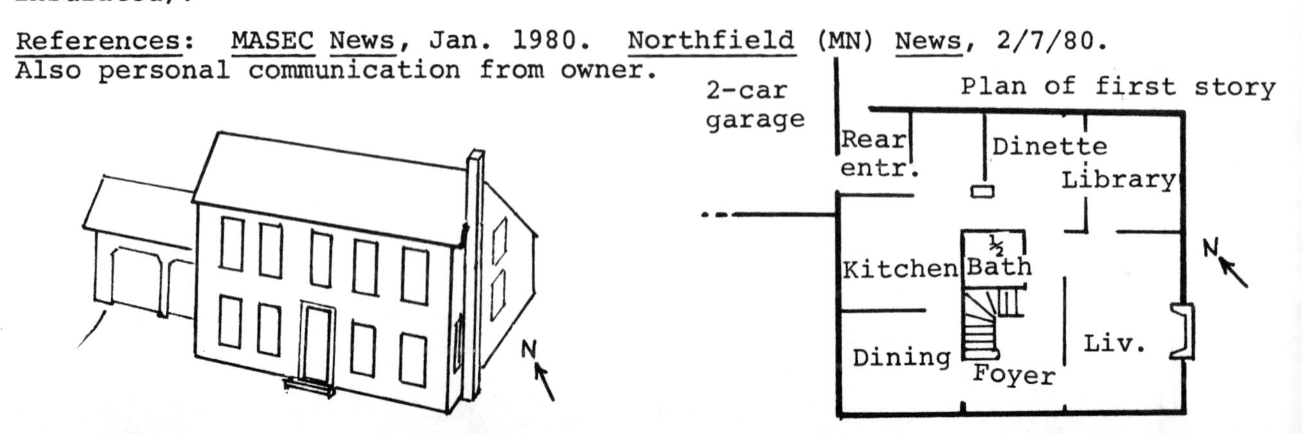

NEW YORK

Tupper Lake, New York
(65 mi. N of Albany)

Bentley House: Mt. Arab district of Tupper Lake, NY 12986. A 8500-DD location. House nearly completed in 1975.

Architect, builder, owner, occupant: R.P. Bentley. Tel.: (518) 359-9300, evenings. Note: This was an experimental house and has been modified from time to time. The improved design has been incorporated in several buildings in the neighborhood (Adirondack Mountain region of New York State). Here I describe only a few of the main features of the design. The combination of features is called "Thermal Efficiency Construction", or TEC, by the originator, R.P. Bentley. Some of the main features are covered by his US Patent 3,969,860. For a copy of his book on the TEC design (see Bibliography item B-237), send a check for $15.75 to R.P. Bentley, Box 786, Tupper Lake, NY 12986.

Some key features of the design:

Exterior wall: 24 in. thick. Filled with fiberglass.

Main floor: (above crawl space): 18 in. thick. Filled with fiberglass.

Ceiling or roof: 18 in. of fiberglass.

Vapor barrier: Used on warm sides of exterior walls, main floor, ceiling.

Construction of exterior wall: Each exterior wall is of stressed skin type. It employs many sheets of plywood or equivalent: 8 ft. x 4 ft. sheets. Thickness may be ½ in. Any given portion of such wall makes use of two such sheets: they are vertical, they are parallel, and they are 24 in. apart. Extending vertically between them are slender built-up beams 8 ft. long, 2 ft. wide. Each employs one or two sheets of plywood strengthened along the long edges by 2x2 wooden strips. Such beams are 4 ft. apart on centers. The big plywood sheets mentioned above are rigidly connected to them by nails or glue. The interior region of each such portion of wall is filled with fiberglass.

The main floor makes use of somewhat similar thick, hollow, rigid, fiberglass-filled structures. The same applies to the ceiling-and-roof system, which employs trusses also.

Vapor barriers (for examples, coatings of low-permeability type) are applied to the warm faces of the warm (inner) plywood sheets. If any moisture penetrates into the fiberglass fillings, this moisture is free to migrate and escape -- i.e., upward, in the case of the vertical portions of walls.

Windows are double-glazed (sometimes triple-glazed). Foundation walls are of concrete 8 in. thick and are not insulated. Eaves are wide. An air-to-air heat-exchanger of special design (about 1 ft. x 2 ft. x 9 ft. high) provides adequately high rate of air-change even in winter months when windows are kept closed.

Warning: This very brief account is far from complete, not entirely clear, and may be inaccurate. For an accurate account, one should refer to the pertinent book B-237.

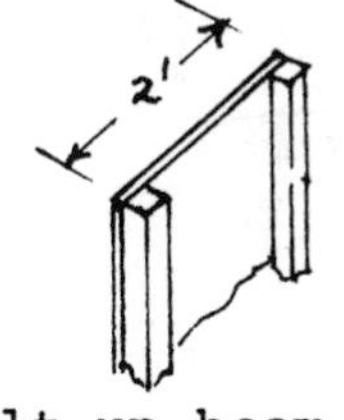

Built-up beam, 2 ft. x 8 ft.

Portion of wall, 8 ft. high, to be filled with fiberglass

VIRGINIA

Springfield, Virginia (15 mi. SW of Washington, DC)

Hart Development Corp. House: 39°N. A 4200-DD location.

Design, construction, and financing: by Hart Development Corp., 5808 River Dr., Lorton, VA 22079. Harry Hart, President.

The two developments: Two developments have been undertaken:

Giles Knoll, Springfield, VA. At or near 7500 Rolling Rd.
Includes 6 houses, representing five models.
All of these houses were sold in 1977 or 1978.

Lee-Brooke, Springfield, VA. At or near 8700 Center Rd.
By 1/1/80 13 houses were under construction. 25 more were planned, to make a total of 38.
Eight models were included.
Typical selling price for house and land: $95,000 to $100,000.

Buildings: These are 1, 1½, and 2-story, 1500-to-2000-sq.-ft., wood-frame houses that employ superinsulation and generally follow the thermal-design features recommended by the Small Homes Council of the University of Illinois and called "Lo-Cal". In all, there are 8 models. Most have 3 bedrooms and an integral 1- or 2-car garage. The entrances to the houses are of air-lock (vestibule) type.

Windows: The window area is about 8% of the floor area. Most of the window area is on the south side of the house. A small fraction is on the north side. Typically there are no windows on the east or west side. Most windows are of casement type. Many of the larger windows are recessed for protection from wind. All of the windows are double glazed.

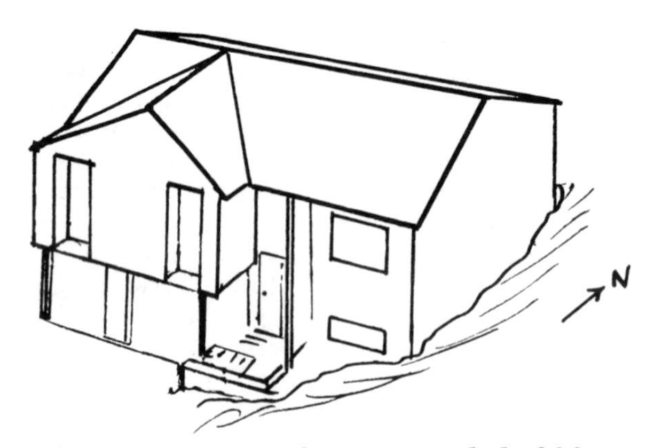

Hart Development Corp. house, Model 200

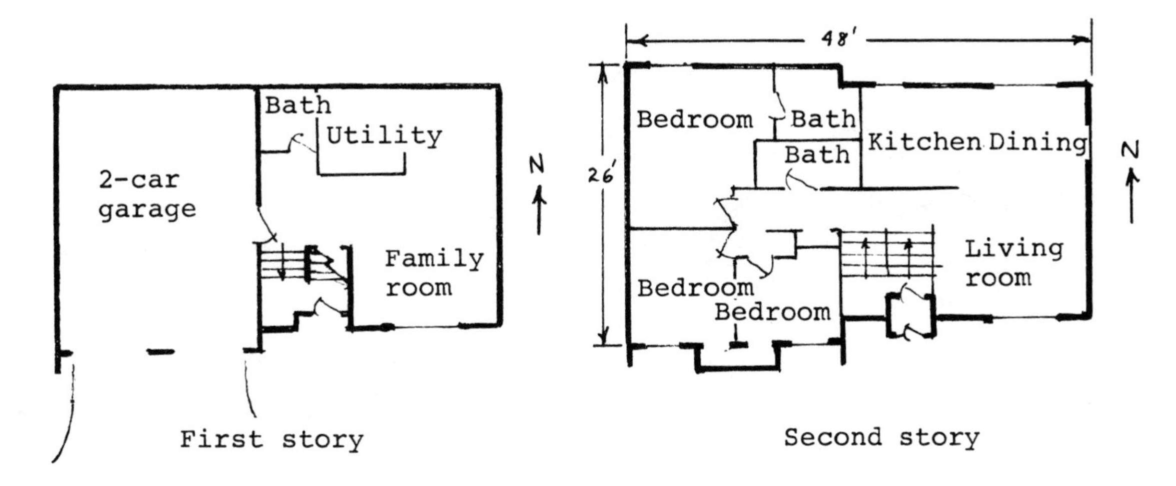

Plan views of the Hart Development Corp. Model 200 house

Wall: There are two sets of 2x4 studs. The outer set has a 16-in. spacing, to match the dimensions of the siding. The inner set has a 24-in. spacing, and no inner stud lines up with an outer stud. Between the two studs there is a 1½-in. space. The outer wall contains 3½ in. of fiberglass (unfaced batts), and the inner wall likewise contains 3½ in. of fiberglass batts; however these latter are faced, on the side toward the rooms, with moisture-proof kraft paper. (The 1½-in. airspace contains nothing except air.) Each of the 3½-in. fiberglass layers provides R-11 and the wall as a whole, including the fiberglass layers, the drywall, and the plywood siding, provides R-24.

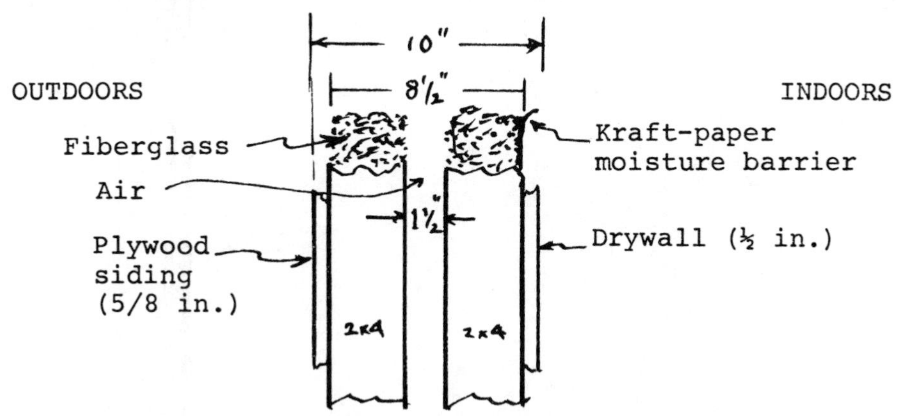

Vertical cross-section of a typical wall of Hart Development Corp. house

Ceiling: This is insulated with 12 in. of fiberglass batts. The combination of this insulation and the ceiling proper provides R-40. The kraft facing of the fiberglass batts is toward the house interior.

Basement wall: This is of poured concrete. The wall is studded on the inside with vertical 2x3s and the space between these is filled with fiberglass; the overall R-value of the insulated wall is R-13.4.

Floor: Floor areas over heated space are not insulated. Floor areas over unheated space are insulated with fiberglass to provide an overall resistance of R-19.

Vapor barrier: No polyethylene vapor barrier is used. The only vapor barriers used are the kraft facings of the fiberglass batts.

Air-change: This has not been measured, but it is estimated that the typical rate of change is 0.5 times per hour. The actual values may vary widely because of the different lifestyles of the occupants.

Humidity: There is no evidence that excessive humidity is a problem. No steps have been taken to control humidity. For example, no air-to-air heat exchangers have been installed.

Auxiliary heat: From 84,000 Btu/hr oil furnace employing a 0.75-gal/hr. nozzle.

Domestic hot water: Heated conventionally by electricity.

Solar heating: The south windows provide considerable heat in winter. However, the solar heating is less important than the all-around excellent insulation.

Cooling in summer: The 26-in. overhang above the south windows excludes much solar radiation in summer. Also it is helpful that there are no east- or west-facing windows. Nevertheless there is need for artificial cooling and a 1½ or 2 ton air conditioner is used.

Performance: Initial performance appears to be good. Six of the houses have been in use throughout one winter. Most used 200 to 250 gal. of fuel oil in that winter.

References: Personal communications from Harry Hart and from W.L. Shick. See also: "Builders Report No. 32", publ. by National Woodwork Mfrs. Assn." "The Energy Puzzle", publ. by Alliance to Save Energy, Sept. 1979, p. 28; also *Builder Magazine* publ. by NAHB, May 21, 1979; also *Money Magazine*, Oct. 1979.

CANADA, SASKATCHEWAN

Regina, Saskatchewan

Pasqua House: 58 Ryan Rd., Regina, Sask. Canada S4S 6V9. 50°N. A 10,600-DD location. Spring of 1979.

Architect: Arthur Allen, in cooperation with Enercon Consultants, Inc. (D.D. Rogoza, Leland Lange, et al).

Builder: Enercon Building Corp.

Building: One-and-a-half-story woodframe house with basement and attached garage. Total living area (not including basement): about 1800 sq. ft. Main story includes kitchen, living, dining, family room, and bathroom. Second story includes four bedrooms and two bathrooms.

Walls: These are of double stud type, are 12 in. thick overall, insulated to R-40. Vapor barriers (0.006-in. polyethylene) are used.

Attic: Insulated to R-52.

Basement: Poured concrete. Walls well insulated. Floor not insulated.

Windows: Double glazed. Equipped with thermal shutters or shades. Total glazed area: 200 sq. ft.

Rate of air-change: 0.3 changes per hour, accomplished by Enercon air-to-air heat-exchanger with about 80% efficiency. (Natural air-change rate: 0.13.)

Solar heating: The large south windows provide much direct-gain solar heating.

Auxiliary heat: From electric heating system of 4 kW capacity. Also a high-efficiency-type fireplace.

Air circulation: Several fans are included (four in all, including the heat-exchanger). One 125-w fan and one 150-w fan operate 50% of the year.

Performance: The auxiliary heat requirement per winter (not counting the above-mentioned fans) is about 7200 kWh, or about 25,000,000 Btu.

Reference: Report "First Annual Pasqua House Report" by Leland Lange, 15 p., March 1980. Publ. by Enercon Consultants Ltd., 2073 Cornwall St., Regina, Sask., Canada S4P 2K6.

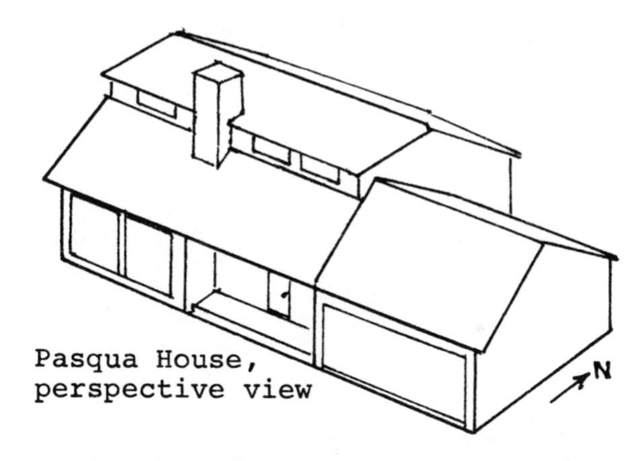

Pasqua House, perspective view

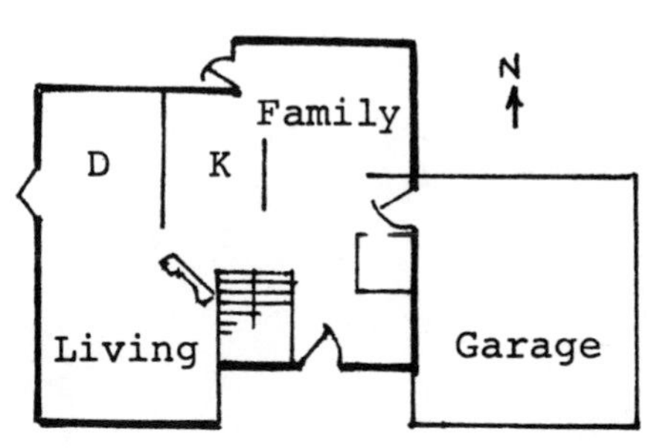

Plan of main story. Windows not shown. Rough drawing only.

Regina, Saskatchewan
(1000 mi. NW of Chicago)

Enercon House of Conserver Type: $50\frac{1}{2}^{\circ}$N.

Designer, builder; Enercon Consultants Ltd. (D.D. Rogoza et al)

Building: This is a general type of house: a superinsulated house. It makes only modest use of solar energy.

Regina, Saskatchewan
(1000 mi. NW of Chicago)

Enercon House of Sun Type: $50\frac{1}{2}^{\circ}$N.

Designer, builder: Enercon Consultants Ltd. (D.D. Rogoza et al)

Building: This is a general type of house: a superinsulated house. It makes much use of passive solar heating.

Note: The main difference between the Sun-type house and Conserver-type house is that the former makes much use of passive solar heating and does not have a conventional heating system.

Saskatoon, Saskatchewan
(1200 mi. NW of Chicago)

Fretz House: RR 5. 53°N. Completed in 1978 or 1979 (?).

Designer, builder, owner, occupant: Peter Fretz.

Building: Two-story woodframe building of superinsulated type. There is a full basement. House walls and ceilings are insulated to R-40 and R-60 respectively. 0.006-in. vapor barriers are used. Walls include two sets of studs. An air-to-air heat-exchanger is used. There is forced-air electric back-up heating. A tightly closing wood-burning stove is included. There is an outside brick chimney at east end and an ell at west end.

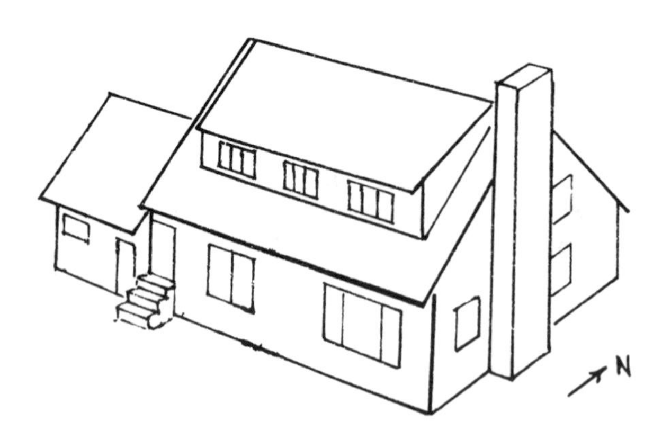

Appendix to Chapter 4

Some Houses That Do Not Quite Qualify

Here I describe briefly some houses that are very well insulated and are of considerable interest but do not quite conform to my definition of a superinsulated house.

MINNESOTA

Northfield, Minnesota (25 mi. S of Minneapolis)

Ivanhoe House; Northfield, MN. 44½°N. An 8000-DD location. Completed in 1977.

Architect: Robert Quanbeck.

Builder: Dallas Haas.

Co-designer and owner: David A. Robinson.

Cost of building (but not land): $61,250. The active solar heating system contributed $6000 to this, and the heat-exchanger contributed $400. Funding: private.

Building: There is a main structure and an auxiliary structure, the latter being north of the former and 12 ft. from it. The two are connected by a north-south enclosed unheated hallway.

Main structure: This has two stories, each 32 ft. x 28 ft. Total floor area: 1800 sq. ft. The upper story includes kitchen and dining area on south, living area on north; at the SE corner there is a screened porch. The entrance to the main structure is at the east end; it makes use of the unheated hallway as a vestibule. The house, of woodframe construction, aims due south.

Auxiliary structure: This has one story and employs slab-on-grade. Plan dimensions: 32 ft. x 21 ft. There is a one-car garage at the east end and a shop at the west end.

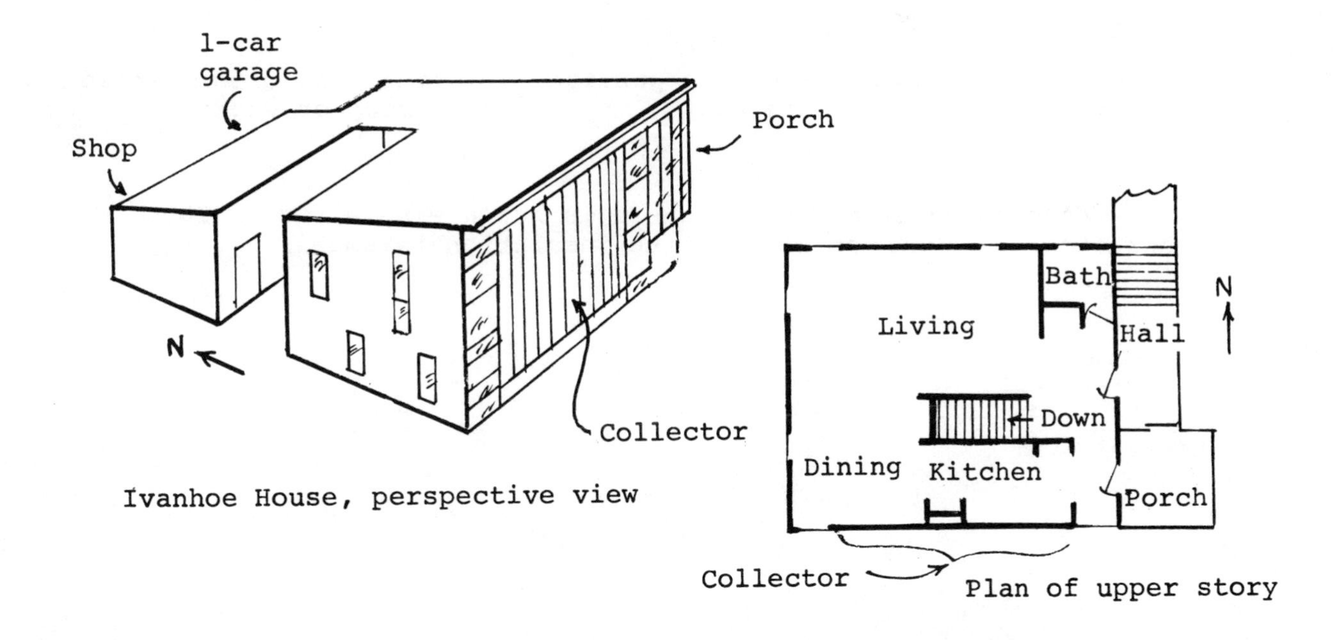

Ivanhoe House, perspective view

Plan of upper story

Windows: The total area of windows on the main structure is 180 sq. ft. Half of this is on the south side. The rest is on west, north, east. (The porch is not included in these figures.) The south windows are double glazed; the others are triple glazed. In south windows there are venetian blinds between the glazings. There are no thermal shutters. The window area is 10% of the floor area.

Window headers are not used. The windows, of casement type, fit directly between the 2x8's 24-in. apart on centers. Also, 2x8's are used at top and bottom of each window to complete the frame. Thus the area not filled with insulation is small.

Wall: The wall is made with 2x8 studs. The 7½-in. space between them is filled with fiberglass. The sheathing just outside the set of studs consists of 1-in. Styrofoam TG (tongue-and-groove). The protective siding consists of 8 ft. x 4 ft. sheets of 9/16-in. plywood. On the room side of the studs there is a 0.004-in. polyethylene vapor barrier protected by 5/8-in. Gypsum board. The overall thickness of the wall is about 10 in. and the nominal overall resistance is R-30.

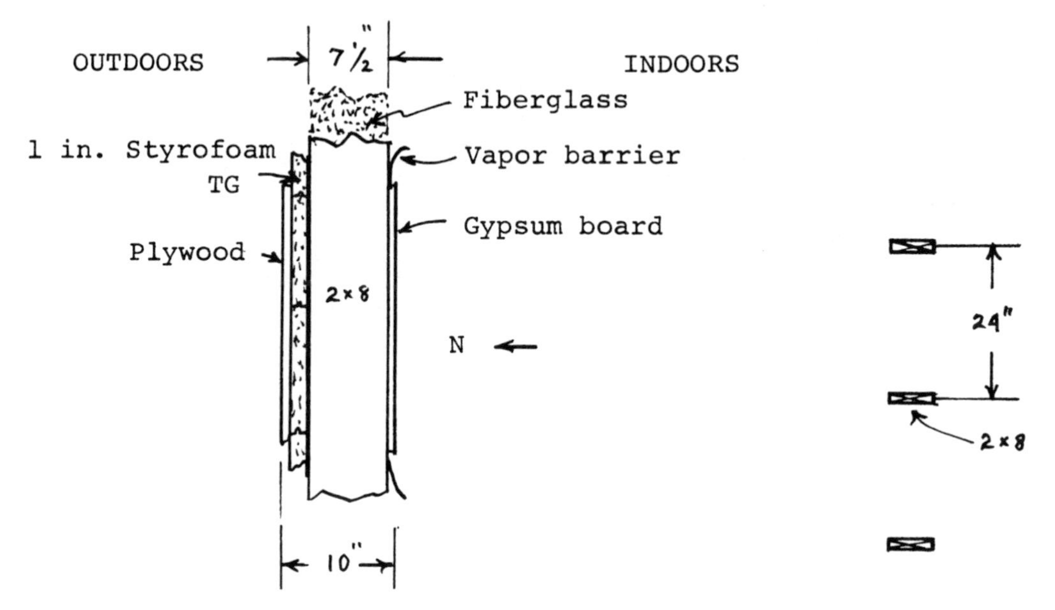

Vertical cross section of a portion of north wall, looking east

Horizontal cross section of a group of studs

At each vertical corner there is but a single stud. Thus here too the area not filled with insulation is small.

The rim joist between upper and lower stories is recessed one inch, and the one-inch space made available is filled with a 1-in. plate of Styrofoam.

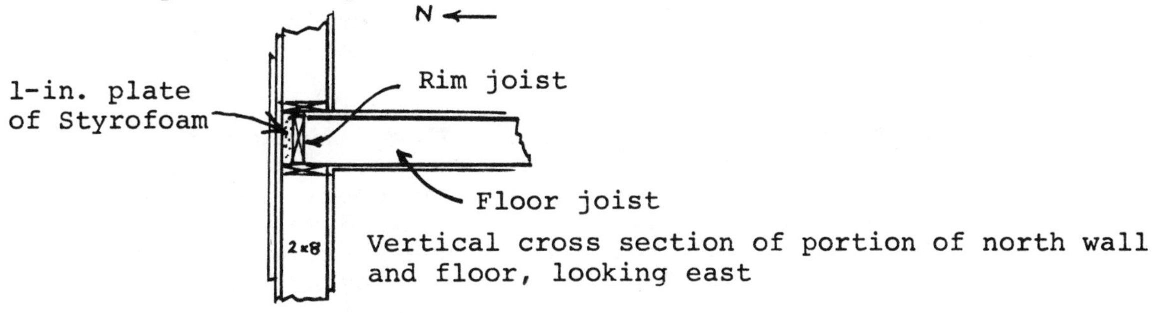

Vertical cross section of portion of north wall and floor, looking east

First story below-grade wall: This is of 8-in. thick concrete blocks insulated on the side toward the room with 4 in. of polystyrene beadboard protected by Gypsum board. Overall resistance of wall as a whole: R-17.

First story floor: Beneath the 4-in. concrete slab there is a 2-in. layer of polystyrene beadboard; near the perimeter the thickness is 4 in.

Upper story ceiling: This includes a 16-in. layer of cellulose, which rests on a 0.004-in. polyethylene vapor barrier. The roof trusses were employed in such manner that the insulation extends to the outer faces of the walls.

Eaves: These are trivial. None projects more than 4 in.

Vapor barrier: 0.004-in. polyethylene vapor barriers are used on walls and upper-story ceiling.

Seals and caulking: Much care was taken to reduce air leakage. Space between window frames and house framing was sealed with urethane foam. Windows are of low-leakage type. Exterior doors have magnetic seals. Electrical and plumbing runs were carefully sealed.

Rate of air-change: Ethane-tracer-gas tests (made with bathroom vent, clothes dryer vent, and heat-exchanger vent open) showed the air-change rate -- on days with 15 mph wind -- to be 0.12 changes per hour. Normally, in winter, an air-to-air heat-exchanger is used and the air-change rate is 0.35 changes/hour.

Humidity: When, in winter, the heat-exchanger is operated two hours each morning and two hours each evening, the humidity remains near-ideal: 40 to 50%. If it is not used (in midwinter), the humidity becomes excessive after a few days.

Heat-exchanger: The air-to-air, counterflow heat-exchanger includes a finned-tube array made by Q-Dot Corp. It employs two fans (together consuming 250 W) which drive the air at 100 cfm. Cost of electricity: About $1/month. The outgoing air provides 85% of the heat needed to warm the incoming air.

Auxiliary heat: 5.5 kW electric furnace with forced-air distribution system. There is also a wood-burning stove.

Domestic hot water: Final heating is electric. Preheating by active-type solar heating system, via coil of copper pipe in storage tank.

Active type solar heating system: 175 sq. ft. water-type vertical collector on south face of house. Copper Roll-Bond panels are used. No antifreeze is used; system is drained at end of day. Glazing: Kalwall Sun-Lite. The 1800-gal. storage tank is situated under hallway and porch at SE. Heat is distributed to the rooms by a finned coil in cold-air-return section of the forced-air system associated with the electric furnace.

Note: The 175-sq.-ft. of collector panels is assisted by adjacent areas (aggregating 150 sq. ft.) of Kalwall Sun-Lite backed by black surface not equipped with water-filled pipes. Such area serves as a slave to the (master) copper panels. The rationale of master-and-slave collection systems is explained in Reference S-235cc.

Warm-air recovery: Warm air from upper part of upper story is circulated by duct and furnace blower to lower story. Return to upper story is via grilles.

Cooling in summer: Windows and doors are opened to facilitate natural ventilation. Also the furnace blower may be used to expel air from upper part of upper story. There are no awnings, but all windows are equipped with venetian blinds. There is no air conditioner.

Performance of main structure in winter: Auxiliary heat need for winter, per DD, is 4100 Btu, i.e., 2.3 Btu/(ft^{2}DD). At night, with outdoor temperature 32°F, room air temperature drops 6 F degrees in 12 hours. The 1/e time constant is thus about 65 hours. About 35% of the heat-loss in winter is made up by intrinsic heat sources and passive solar heating. Another 40% is by the active solar system. If the rest of the heat is supplied by the electric furnace, about 2500 kWh of electricity is required, i.e., about $100 worth at 4¢/kWh. (The figure is much lower if much use is made of the wood-burning stove.)

References: Various reports by David A. Robinson: R-190, R-192, R-194, R-196.

Comment: The owner feels that the active-type collector and the storage tank may be too large. He feels that it would have been advantageous to have used (a) a smaller active collector, (b) a greater passive-collection area, (c) more insulation in walls of house, and (d) thermal shutters or shades on windows at night. The special type of collector used was not intended, but was a consequence of unexpected problems with the collector as initially installed and an improvised solution to those problems.

NEW MEXICO

Superinsulated Dome

I understand from J. Baldwin that he made some tests on a superinsulated 30-ft.-diameter dome in New Mexico and found that the indoor temperature remained comfortable -- thanks to inherent heat sources -- until the outdoor temperature dropped to 0°F. Even in the coldest weather (and with ½ air change per hour), the only auxiliary heat needed was that from an ordinary Coleman lamp. The dome had been insulated carefully with urethane foam and fenestration and infiltration had been carefully controlled.

NEW YORK

Edinburg, New York (200 mi. N of NY City)

Brownell House: Edinburg, NY 12134. An 8300-DD location. 1977.

Designer, builder: B.R. Brownell of Adirondack Alternative Energy. Tel.: (518) 863-4338.
Note: Brownell, a pioneer in superinsulated house design, has designed and/or built scores of very well insulated, well sealed houses in the period 1977 - 1980. Here I describe one of his earliest houses; I mention only a few of the main features of the design. Detailed information may be obtained from his company Adirondack Alternative Energy (address given above).

Building: This is a 2300-sq.-ft. house (two stories and basement) that is excellently insulated and sealed. Insulation in walls and roof: 4 in. of Thermax; specifically, a series of two sheets each 2 in. thick. Overall R-value: 30 to 36. The aluminum-foil faces of the Thermax plates serve as vapor barriers; also the (isocyanurate) foam itself does not transmit or accept water and accordingly problems of moisture penetration and condensation are avoided. Because Thermax provides 2 to 2½ times as much R-value per inch as fiberglass provides, walls do not need to be 11 or more inches thick (as in many other kinds of very well insulated houses) and preemption of valuable space is avoided. (Note: fiberglass has important advantages of fire resistance and low cost.) Window area: 390 sq. ft. Most of this (250 sq. ft.) faces south. The designer's intent is that much solar energy be collected and stored in the sand bed discussed below.

Sand bed: A major feature of the house is a 115-ton sand bed that is 2 ft. thick (some newer houses have sand beds 4 ft. thick). It extends beneath the entire area of the concrete floor slab of the basement. Buried in the sand are main ducts and junior ducts, the latter being a few feet apart on centers. Beneath the sand bed there is a 4-in. layer of foam (two sheets of 2-in. Thermax) and below this there is a vapor barrier of polyethylene.

Chimney structure: Near the center of the building there is a vertical 20-ton chimney structure that contains a flue (for wood stove, e.g.) and also vertical ducts for carrying warm air from upper part of clerestory to ducts in sand bed. Two blowers at base of chimney structure provide an airflow of 1000 cfm.

Auxiliary heat needed: Said to be equivalent to 230 to 390 gal. of oil. (But in typical, more recent, Brownell houses the requirement is only about 110 gal.) A wood stove may be used, and/or electrical heaters.

Reference: J-400, a 1979, 72-p. report by R.F. Jones et al of Brookhaven National Laboratory. See also an article on p. 25 of *Alternative Sources of Energy* #40, Nov.-Dec. 1979.

Upton, L.I., New York
(70 mi. E of NY City)

Brookhaven House: Upton, L.I., NY. On grounds of Brookhaven National Laboratory. 41°N. Construction started Jan. 1980.

Architect: Total Environment Action, Inc.; P. Pietz, L. Heschong.

Owner: Brookhaven National Laboratory (of DOE).

Builder: Thermal Comfort, Inc.

Cost: In $80,000 to $100,000 range; perhaps less if built routinely instead of for demonstration. Funding: by DOE.

Building: Two-story, three-bedroom, 1½ bath, woodframe house with attached greenhouse, partial basement, crawl space, detached two-car garage. Floor area of first and second stories, incl. greenhouse: 2350 sq. ft. Air-lock type exterior doors.

Windows: Most are of casement type and are triple glazed. Window areas (not including greenhouse) on south, west, north, east are: 60, 30, 35, 30 sq. ft.

Greenhouse: Total floor area: 220 sq. ft. East portion is one-story high, west portion is two stories high. Glazing is double. Construction is of conventional "kit" type. Wall between greenhouse and house proper is massive: of paving bricks. East end of greenhouse is opaque and insulated; no glass here.

Wall: A typical wall has a single row of 2x6 studs 24 in. apart on centers with 5½ in. of fiberglass insulation between and, on the outside of the wall, 1 in. of Styrofoam TG. On side toward room there is a 0.006-in. polyethylene vapor barrier. Overall R-value of wall: R-27. There are no electrical outlet boxes in the outer walls; instead, surface raceways (on wall faces toward center of room) are provided.

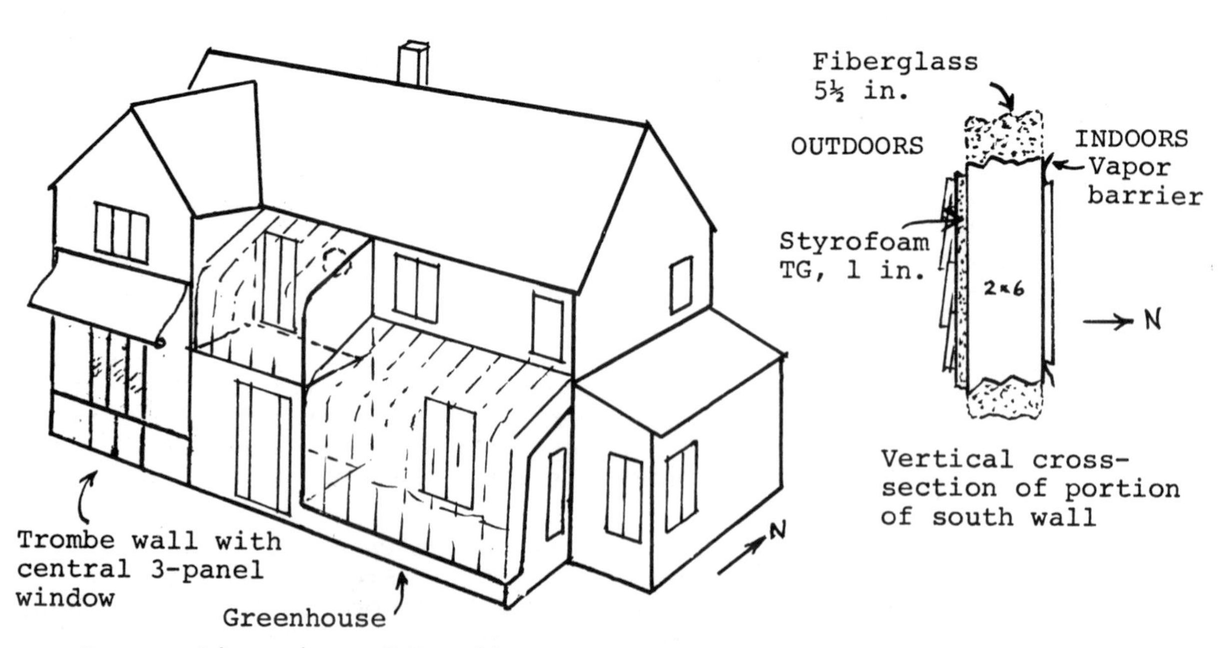

Perspective view of Brookhaven House

Vertical cross-section of portion of south wall

Sills: These have only 1 or 2 in. of insulation. The same applies to the headers.

Floor perimeter: Just beneath the perimeter of the first-story floor (and likewise the second-story floor) there is an 11½-in.-thick band of fiberglass.

Attic: The attic floor is insulated with 12 in. of fiberglass. There are continuous vents along the soffits and the roof ridge. Overall R-value of roof and attic floor: R-38.

Partial basement: This is under the east end of house, i.e., under kitchen and family room. The outside of each basement wall is insulated with 2 in. of Styrofoam. Under the rest of the house proper there is a crawl space. The floor of the crawl space consists of a 2-in. concrete slab which rests on 2 in. of Styrofoam.

Circulation of air: Hot air from greenhouse may be allowed to circulate passively into the adjoining rooms. Grilles permit warm air in the first-story rooms to flow up into the second-story rooms.

Trombe wall: This wall, of paving bricks, is at the west end of the south wall. The south face of the wall is insulated with three layers of float glass. There is a central rectangular aperture (window, 5½ ft. x 4 ft.) in the center of the wall. Wall thickness: 8 in.

Vapor barriers: These are of 0.006-in. polyethylene and are applied to the inner (warm) surfaces of all external walls and second-story ceiling.

Dehumidifier: None.

Air-to-air heat-exchanger: None.

Domestic hot water: Heated by conventional electric heater. (Solar option).

Auxiliary heat: Wood stove (backed by massive brick wall). Also gas furnace.

Cooling in summer: Within the greenhouse, interior roll-down shades are used. The Trombe wall is shaded by an exterior, roll-down, swing-out awning. There is a 1000-cfm blower that, on hot summer days, drives hot air in upper part of greenhouse into the attic space whence it flows, via soffit and ridge vents, to the outdoors. The blower can also be used to serve the entire house.

Performance: It is expected that the house will require only a modest amount of auxiliary heat per winter -- an amount equivalent to 130 gal. of oil. If the furnace is not used, the wood-burning stove can be used and will consume about one cord of wood.

Is this a superinsulated house? In many respects it is. However, it fails to conform to my definition inasmuch as it has (1) deliberately added thermal mass, (2) a large area for solar collection -- greenhouse and Trombe wall, (3) less than 7 in. of insulation in wall, and (4) a furnace. More properly it may be called an indirect gain, passively solar heated house.

References: Solar Age, Jan. 1980, p. 56. See also "Design of Residential Buildings Utilizing Thermal Storage", Report No. DOE/TIC-10143, NTIS, $9.50. Also see a 32-p. 1979 brochure prepared by Total Environmental Action, Inc.

Chapter 5
SUPERINSULATED HOUSES: DISCUSSIONS OF COMPONENTS

INTRODUCTION

Here I discuss each component, or each element, of the superinsulated house. (The performance of the house is discussed in the following chapter.)

INTRINSIC HEAT

This is discussed in Chapter 2.

DESIRED INDOOR CONDITIONS

Occupants may want the rooms to be at 70°F.

A lower temperature may be appropriate if they are exercising or are warmly dressed -- or if the walls, floors, etc., are approximately as warm as the room air, and if there are no air currents (no drafts), and if the humidity is reasonably high, such as 40% or 50%. In summary, there are three reasons why it is permissible, in a superinsulated house, to set the thermostat especially low: warm walls, absence of drafts, and high humidity.

WINDOWS

The special criteria governing locations and areas of windows of a superinsulated house are these:

The area of south windows should be moderate: large enough to admit much solar radiation on sunny day, but not so large that heat-loss on cold nights is very large or heat-gain on hot summer days is very large. Typically, the area should be about 5 to 10% of the floor area. Many books and articles listed in the bibliography deal with this question. See especially S-185, L-195.

The area on each other side of the building should be much less. Say 1% or 2% of the floor area. Sometimes it is feasible to have no window on one or two of these other sides.

South windows should be double glazed -- or triple glazed in very cold regions. Other windows should be double, triple, or quadruple glazed. The third (or third and fourth) glazing may be of informal inexpensive type; e.g., a sheet of transparent plastic employed just during the winter. (Note: Typical R-values of single, double, and triple glazed windows are about 0.9, 1.8, and 2.7 respectively.)

Use of thermal shutters or shades may be desirable.

The windows should be almost perfectly airtight. I am told that casement windows are usually tighter than double-hung windows.

The windows should provide adequate daylight illumination and adequate view in several directions.

Some windows should be designed so that they can be opened, as for ventilation in summer, or for venting excessive humidity in winter, or for permitting occupants to escape if fire breaks out and doors are not negotiable.

To increase the amount of solar radiation entering the house via its modest-size south windows, one may equip these with near-horizontal reflectors installed outdoors at window-sill level. Details are given in S-235cc.

East and west windows may be equipped with vertical reflectors. See S-235cc.

In the period Aug. 15 - Oct. 31, half-shutters may be used to cover the lower halves of the south windows on sunny days, to exclude sun's rays that are not blocked by the eaves. Thus overheating of the rooms on those days may be avoided.

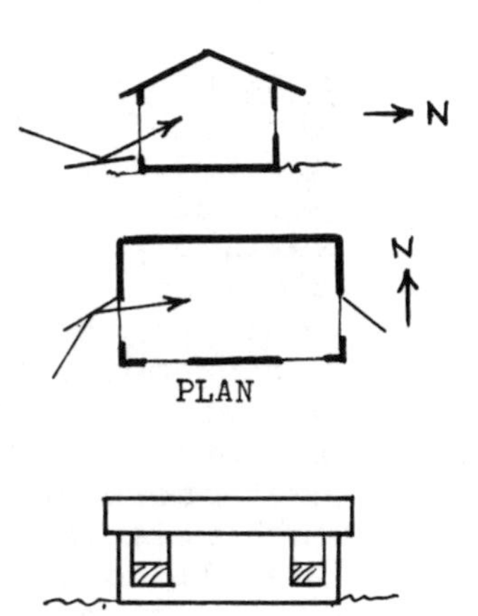

Proposal for Use of Larger Area of South Windows

It seems to me that, in summer, awnings or shutters to exclude solar radiation should be used. They are effective against all kinds of radiation: direct, diffuse, and reflected-from-ground.

Also it seems to me that, in winter, thermal shutters or shades should be used. They cut nighttime heat-loss-through-windows by 50% to 85%.

If these improvements are made, use of a larger window area is advisable. See L-195. Wintertime intake of solar radiation is increased, and view is improved. Also, (1) heat-loss at night is small, and (2) summertime intake of solar energy is decreased.

WALLS

Walls must be well insulated -- say to R-30 or R-40, for typical locations in the northern half of USA. Usually this requires about 8 or 9 inches of fiberglass of cellulose fiber. Insulating sheathing (of Styrofoam of Thermax, for example) may serve in lieu of a portion of the fiberglass or cellulose fiber.

Walls must also be practically airtight. Vapor barriers (discussed on a later page) can insure this. The integrity of such barriers must not be spoiled by holes, gaps, etc., for electric outlet boxes, pipes, etc.

Walls must be strong, of attractive appearance, easy to construct.

They must not constitute a significant fire hazard. Fiberglass itself is non-flammable; but an asphaltic kraft paper backing (if any) can burn. Most foam-type insulating boards include fire inhibitor, but under severe conditions may burn; when they burn they emit poisonous gases. Cellulose fiber, unless it contains the proper amount of proper fire inhibitor, can be a serious fire risk.

Note concerning thickness of insulation: In several recent articles (R-192, R-194) D.A. Robinson has developed approximate formulas for the optimum thickness, or optimum R-value, of walls -- assuming various values of cost of insulation per unit R-value and cost of auxiliary heat per Btu. He finds that, if certain simplifying assumptions are valid, the optimum R-value is proportional to the square root of: the second of these quantities divided by the first. Thus, for example, if the cost of auxiliary heat increases, in a certain period, to four times its present cost, one should use enough insulation to provide twice as great R-value. Or, if cost of auxiliary heat increases 2%, the wall R-value should be increased 1%. However, such rule is hard to apply if one must take into account the cost of installing the insulation, cost of installing additional studs, cost of the space preempted by the insulation, inflation, interest rates, governmental tax credits, and the owner's income tax bracket.

One of Robinson's conclusions is: "The general implication is that on a life-cycle cost basis most homes are underinsulated by a factor of two even when heated with inexpensive energy (natural gas at $2 per million cubic feet)." If expensive fuel is used, the underinsulation factor presumably far exceeds 2.

Note concerning gaps in insulation: The actual R-value of a wall may be much less than one would assume from the published R-value of the insulating plates used. Unintended and unsuspected cracks or gaps in the insulation -- due to poor initial installation or subsequent shrinking -- often occur, permitting leakage of air and heat. In one series of tests reported on a wall containing insulating boards of Styrofoam or other foam-type materials, the overall R-value was found to be only 2/3 of the naively calculated value. (See "Study Analyzing the Effects of Sheathing Products on Wall Heat Losses Caused by Air Infiltration", by V.G. DeNunzio and E.G. Strass of Simplex Industries, Nov. 10, 1978).

Note concerning use of thinner insulation on south side of house: Some tests have shown that, in a well-insulated house in a sunny region, the thickness of the insulation in the south wall should be only about half that in the other walls. If the other walls are R-19, and the R-value of the south wall is only half this, the net heat-loss through the south wall may be less than if thicker insulation is used. See Bibl. item A-256. I do not know for what range of insulation thicknesses, or what range of climates, this conclusion is valid. Also, the key question may not pertain to average heat-loss throughout the winter, but may pertain, rather, to this more critical matter: heat-loss during a long, cold, overcast period. It is the size of this latter heat-loss that may determine whether a very well insulated house can get by without a furnace. In a long cold overcast period, it is best if all of the walls of the house have approximately equal insulation.

Note concerning cellulose fiber: This material is cheap, readily available, has high R-value, and is easy to install. If properly treated with boric acid and borax (in about the proportions 2-to-1), it has considerable resistance to fire; yet under severe conditions it will burn. If other inhibitors are used, or the amounts used are inadequate, or if after some years the inhibitor has been leached away by seeping water, one may need to be concerned about fire hazard.

(Reference: article by R.P. Benedetti in May, 1978, "Fire Journal".)

Note concerning fire hazard in an especially tightly sealed house: Here are quotations from an article by J.C. Degenkolb in the May, 1978, Fire Journal:

> In a certain ASHRAE standard on insulation "..you will find no mention of fire safety and how the burning characteristics of the building may be modified (by the proposed insulation) and, rest assured, they will be!"
>
> "The fact is that heat, smoke, and combustion gases can't escape from an insulated and sealed building -- and that is what energy conservation is supposed to provide. If we have a tight, energy-conserved, building, an incipient fire can go flashover much more rapidly."
>
> "There is the distinct possibility that some types of thermal insulation may cause a shorting out of knob and tube wiring due to the presence of flame-retardant salts and moisture."
>
> "Even the best of these flame-retardant treated cellulosics will begin to smolder when heated to approximately 450°F. When smoldering once begins, it is often difficult to extinguish."
>
> Unless installers of insulation proceed with great knowledge and care, "we are building for a rash of fire deaths..."

Installation, in a superinsulated house, of one -- or several -- smoke detectors and alarms should help considerably in reducing deaths from fire.

Note concerning upward migration of moisture between the two walls: If there are, in each wall, two sets of studs, and there is some freedom for moisture here to migrate upward and to vent into the attic space and thence to outdoors, this can be helpful. The double-wall construction may offer this constructive option.

Note concerning moisture in outer portions of fiberglass insulation: I have been told by Bruce Brownell that he has found that, in winter, in certain well-insulated houses in the Adirondack region of New York State, there is often much moisture in the outer portions of the fiberglass insulation in the exterior walls. Despite the fact that the houses in question were well insulated and were provided with vapor barriers, a substantial fraction of the fiberglass (the coldest portion) was found to be moist, and accordingly the effective R-value was far below the nominal value. Most vapor barriers, even if installed with reasonable care, contain holes, and there may also be gaps at the edges. Whatever moisture finds its way past the barrier may condense in the colder regions of the fiberglass.

To avoid this trouble, Brownell has employed, in the houses he designed, foil-faced foam rather than fiberglass. For example, he often employs two sheets of 2-in. Thermax in series. The aluminum foil facings serve as vapor barrier and in any case the foam itself is waterproof. Cracks between adjacent Thermax sheets are carefully sealed.

Sills, headers, window frames, etc.: Great care should be taken to insure that each of these components is provided with at least a moderate amount of insulation, e.g., enough Thermax, Styrofoam, or fiberglass to provide at least R-5 and preferably more. Some designers favor use of box-type headers; see S-185, p. 42.

How should one apportion money between insulation for different components of a house? This question has been analyzed by D.A. Robinson (R-192, R-194). Where certain simplifying assumptions are applicable, a square-root rule applies. Example: if the per-square-foot, per-unit-of-R-value cost of window insulation is four times that of wall insulation, one should plan on having, for the walls, twice (not four times) the R-value of the window insulation. If ceiling insulation were to cost 2% more (in these same units) than wall insulation, the R-value chosen for the ceiling should be 1% less than the value chosen for the walls. (Note: $\sqrt{1.02} \cong 1.01$.)

Thin wall proposed by D.G. Fuller: Reasonably high R-value can be achieved in a thin wall by using a scheme proposed by D.G. Fuller of Environmental Design Alternatives of Kent, Ohio. There is but a single row of 2x4 studs, and most of the space between them is filled by a 3-in.-thick plate of urethane foam. A 3/4-in. sheet of Thermax covers the entire outer face of the wall and, besides improving the wall insulation generally, it provides significant (R-6) insulation for the studs themselves. The Thermax is protected by a sheet of plywood. The urethane and Thermax together serve as vapor barrier. The R-value of the wall as a whole is about 30.

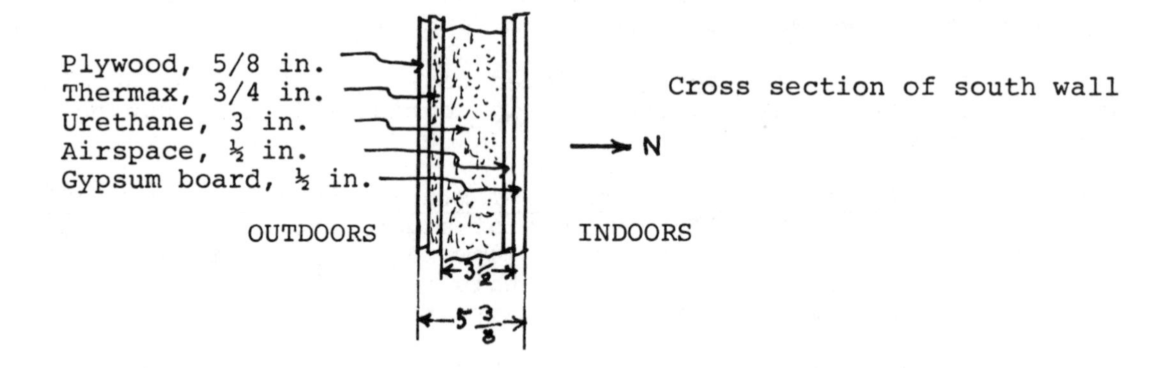

Cross section of south wall

ATTIC AND ROOF

Here, typically, 10 to 12 inches of fiberglass is used. The typical R-value is about 35 to 40. Some designers use R-60.

It is important to extend the insulation outward far enough to cover the areas above the walls.

Also it is important that there be no gaps in the vapor barriers. Any penetrations, as for ventilation pipes or chimneys, should be well sealed.

Often, full-length vents along the roof ridge are provided. Or large vents at the end gables. Often full-length soffit vents are used.

FLOOR

Heavy insulation should be provided to underside of the first-story floor unless the foundation walls are well insulated. R-29 to R-36 is typical, and a vapor barrier is used.

FOUNDATION WALL, BASEMENT, CRAWL SPACE

Usually the concrete foundation wall is insulated on the outside with 1 or 2 in. of Styrofoam. If there is a basement, it may have a 4-in. concrete floor beneath which is sand or gravel.

AIR-LOCK ENTRY

In the first edition of this book, I wrote:

> "Outside entrances should be of air-lock (vestibule) type. In some cases the function of a vestibule may be served by a small adjacent greenhouse or by an attached garage. Some designers make the vestibules large enough so that they may serve several purposes; closets, utility rooms, etc., may be accommodated."

But I have recently been persuaded by Bruce Brownell, a pioneer of superinsulated houses, that, in many situations, inclusion of a vestibule is not justified. Some added expense, or preemption of space, is entailed. In mid-winter the occupants leave or enter the house only a few times a day, and the brief bursts of fresh, low-humidity air may be welcome. Anyway, most occupants may be careless and may leave the inner door open at all times -- in which case the air-lock function of the vestibule is defeated.

Suggestion concerning use of smaller door: Typical front doors are made large enough to admit sofas, pianos, etc. If the doors were much smaller, the need for air-locks (i.e., the need for vestibules) would be much less. Could one design a two-area door: a door-within-a-door? The larger door would be large enough to admit pianos etc., but the smaller door would be much smaller: it might be only 16 in. wide and 70 in. high. In winter, only the smaller door would be used, ordinarily; thus the amount of heat-loss would be reduced about 60%, and a vestibule would be superfluous.

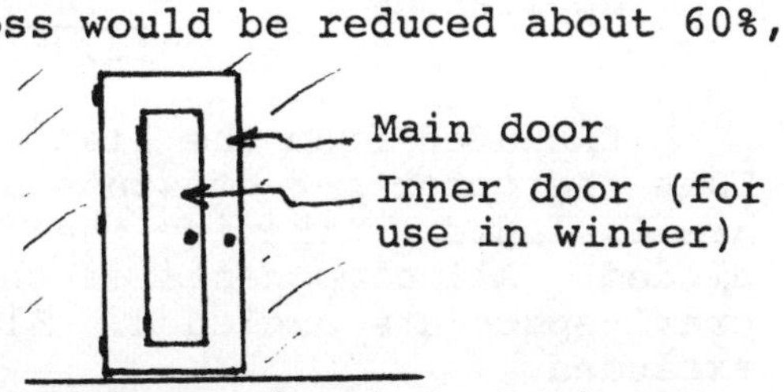

Main door

Inner door (for use in winter)

THERMAL MASS

Part of my definition of a superinsulated house is that the house contains no added thermal mass. Its normal, natural mass suffices.

What controls the rate at which a house cools down during a cold period in winter when no heat is supplied to the house is the product of thermal mass and the thermal resistance of walls, roof, etc. Obviously, the designer wants the rate of cool-down to be low, and accordingly he wants this product to be large.

Another reason for wanting the product to be large is to minimize the rate of heat-up when energy is added. If the house warms up only a few degrees when 50,000 Btu of solar energy is received, the occupants will not open doors or windows; they are content to have the energy remain in the house, and the energy will be useful in keeping the house warm during the coming night.

In typical direct-gain solar houses, the thermal resistance values are fairly low, and accordingly it is necessary, on cold nights, to have very large thermal mass. If the area of south windows is very large, and if these windows are single glazed and are not equipped at night with thermal shutters or shades, the rate of heat-loss is very high and it may be imperative to have large thermal mass.

But in a superinsulated house the insulation is superb and consequently it is permissible for the thermal mass to be small -- of the order of 15,000 to 30,000 lb. of wood, gypsum board, etc.

Effective thermal mass: The effective thermal mass of a superinsulated building is the mass of components and contents that are fairly well or very well insulated from the outdoors and are thermally coupled to the rooms.

Saskatchewan Conservation House: Here, the wall components, floor components, ceiling components, etc., that are thermally isolated from outdoors and thermally coupled to the room air have a mass of 22,000 lb. and a heat-storage capacity of about 10,800 Btu/^{o}F, according to Besant et al (B-250, B-251, B-251c). A list of the contributions by components follows.

	Mass		Thermal capacity		
Component	kg	lb	MJ/^{o}C	Btu/^{o}F	Percent
Gypsum board	5006	11,000	5.5	2,900	27%
Wood					
studs	2072	4,600	3.9	2,050	63%
flooring	2034	4,500	3.9	2,050	
floor joists	2286	5,000	4.3	2,260	
2nd story ceiling joists	453	1,000	0.9	470	
Furniture and appliances	1000	2,200	2.0	1,054	10%
	12,951	28,300	20.5	10,800	100%

Excluded from the list are all components outside the insulation. Thus the outer set of studs and all material farther out than these are excluded. Likewise the floor joists of the first story are excluded. All components of the large (unheated) attic and (unheated) crawl-space are excluded. Also the large steel storage tank is excluded.

Note that about 2/3 of the effective thermal mass is attributable to wood and about 1/4 is attributable to gypsum board.

How to increase thermal mass: If a designer is resolved to increase the thermal mass of a superinsulated house he is designing, he may wish to use two layers of gypsum board, rather than just one. The second layer, behind the layer that is visible in the room, may be of gypsum boards that are partly broken (rejects) and in a sense are free (suggested by R.W. Besant). Also, the builder could make some of the partition walls of bricks or concrete blocks. Of course, water-filled containers could be used.

VAPOR BARRIER

Vapor barriers are used in superinsulated houses for two important reasons: (1) to prevent water vapor from reaching the cold parts of the walls (colder parts of the insulation, colder framing members, etc.) and condensing there, forming pools of water or ice, degrading the performance of the insulation and perhaps eventually leading to rotting of the wooden studs, sills, etc., and (2) to prevent in-leak or out-leak of air and prevent the associated heat-loss.

Materials: Most vapor barriers are of polyethylene, e.g., a polyethylene sheet 0.006-in. thick. Sometimes thinner (0.004 in.) sheets are used. The aluminum-foil faces of Thermax constitute moisture barriers. The aluminum-foil backing of certain fiberglass batts is a vapor barrier. Some asphaltic kraft paper backings of fiberglass batts or blankets is fairly effective as a vapor barrier.

Acetate-type wall papers are somewhat effective, I understand.

Can paint serve as a vapor barrier? I understand that some kinds can. For example, a paint called Insul-Aid and sold by Glidden Durkee Division of SCM Corp. is said to be very effective. Its permeance is said to be very low. It may be procured from the main company, at 3rd and Bern Sts., Reading, PA 19603, or, in New England, from a branch at 155 Main St., Stoneham, MA 02180. Some enamel paints are said to be effective, especially if two coats are used. To be avoided are most of the common kinds of paints -- Latex paints especially.

Location of application: In cold climates, vapor barriers are usually placed on the warm sides of walls, ceilings, etc. However, an alternative location is permissible and, in some instances, preferable: a location such that one third of the insulation is on the warm side of the barrier and two thirds is on the cold side. So situated, the barrier will always be warm enough so that no moisture will condense on it; and the builder is now free to install pipes, wires, outlet boxes, etc. within the shallow region on the warm side of the barrier with no risk of cutting holes in it.

In very hot climates the vapor barrier may be placed on the outer side of the wall, to help exclude moisture threatening from outside on very hot and humid days when the rooms themselves are kept cool.

Seals: Sealing the vapor barriers along their edges is important. Normally the builder overlaps adjacent barriers (sheets) and seals the overlap. The overlap should occur where there is a solid backing, e.g., immediately in front of a stud or header. Sealing compounds are usually used; I am told that there is available an "acoustical" sealant that is especially to be recommended because it never becomes hard and brittle.

Backing: I understand that it is important that the vapor barrier have a reasonably firm backing -- which will prevent it from moving, bellying, etc., when there is a very high wind. If it moves or bellies, gaps may soon arise at the edges, and possibly the sheet may tear.

Is it harmful to employ two vapor barriers -- one on each side of a 9-in. layer of fiberglass or cellulose fiber? If some moisture finds its way in between the two barriers, will it stay there permanently and lead to rotting of framing members? Or will it somehow find its way out to outdoors? Would it be helpful to provide a path whereby such moisture can migrate parallel to the barriers and find an exit elsewhere, e.g., at the eaves? I do not know the answer.

Avoiding holes and gaps: All holes and gaps in vapor barriers must be sealed. There may be gaps at locations where pipes, or wires, of chimneys pass through the barriers. These gaps must be found and sealed. The problem is simplified if the pipes, wires, etc., can be routed so as not to encounter the vapor barriers. For example, the builder may arrange for there to be no electrical wires within any outside wall; outlet boxes may be limited to partition walls -- or may be provided by means of wiring strips affixed to the warm faces of the outer wall, i.e., strips everywhere within the space defined by the vapor barriers.

One designer has told me: "Integrity (of the vapor barrier) is more important that low permeance to H_2O."

I have also been warned that there is risk in using big plates of polystyrene foam or certain other materials as vapor barrier. Perhaps such plates will shrink or warp or for other reason change in dimensions, with the results that cracks may open up along the edges. It is much safer to use large flexible sheets of polyethylene.

Reminder re importance: Vapor barriers are of first importance. The builder of a superinsulated house should recognize the need for giving much thought, and close supervision, to installing the barrier and making sure that its moisture-proof integrity is preserved. In typical houses of some years ago, vapor barriers were relatively unimportant; abundant leakage of air would solve nearly all moisture problems. But to the success of a superinsulated house, vapor barriers are crucial.

"Moisture Condensation" pamphlet by University of Illinois: The general problem of moisture condensation within walls, floors, etc., of houses is discussed in an 8-p. pamphlet (Pamphlet F6.2, $0.25) published in 1975 by the Small Homes Council of the University of Illinois. A wide variety of solutions are described.

"Predicting Moisture Condensation in Walls": 14-p. article by Dan Lewis, Winslow Fuller, and Paul Sullivan in the Sept. 1980 Design Note 5 by the Northeast Solar Energy Center, 470 Atlantic Ave., Boston, MA 02110. This excellent article, based in part on Chapters 19 and 20 of the ASHRAE 1977 Handbook of Fundamentals, explains why exceptionally high integrity of vapor barriers is required if (a) the weather is very cold, (b) there is much moisture in the house, (c) R-values of walls are high, (d) the rate of infiltration is low, and (e) the outer layers of the outer walls are relatively moisture-proof.

RATE OF AIR-CHANGE

A perennial topic of conversation among proponents and critics of superinsulation is rate of air-change. To make the rate very small is an obvious goal, inasmuch as air-change may be responsible for 25% to 50% of the heat-loss of a very tight, very well insulated house; if such house is to get by with no furnace, clamping down on air-change is a must. But alas, if the rate of air-change is very low, spectres of stuffiness, kitchen smells, bathroom smells, cigarette smells, excessive moisture, and poisonous-gas build-up are sure to arise. In clamping down on air-change, the designers want to "run it close" but not "too close". There can be no simple criterion. Different minimum-rates will apply depending on family size, family habits, and other circumstances.

Some actual values: In a number of superinsulated houses, such as the Saskatchewan Conservation House and various houses built by Enercon Consultants Ltd., the air-change rate is as low as 0.05 changes per hour if all windows are shut, ventilating fans and air-to-air heat exchangers are off. Various other houses have 0.1 to 0.3 changes per hour.

Consensus as to acceptable minimum rate: From talking with experienced designers and reading their reports, I gather that:

0.2 changes per hour may be acceptable if the family is small, if there are no unusual sources of moisture or smell, if exhaust fans in kitchen and bathroom are operated briefly when appropriate, and shower baths are kept brief. (See S-185, p. 110).

0.3 changes per hour may be the best general goal.

0.5 changes per hour may be comfortably adequate and perhaps slightly excessive.

How much sensible heat is lost by air-change? A cubic foot of 70F air weights 0.077 lb. Its specific heat is 0.24 Btu/(lb. ^{o}F). If a cubic foot of 70F air is replaced by a cubic foot of 30F air, the amount of sensible heat lost is 0.077 x 0.24 x 40 = 0.74 Btu. If all the air in a 10,000-cubic-foot house (say: a house with 1200 sq. ft. floor area and 8-ft. ceiling height) is replaced each hour with 30F air, the heat loss is 0.74 x 10,000 = 7400 Btu/hr. Thus the loss in a 24-hr. period is 24 x 7400 = about 180,000 Btu. If the outdoor temperature is -10F, the loss is twice as great: 360,000 Btu.

These losses are large! They motivate the designer to reduce the air-change rate as far as permissible. (Some years ago I would have said "as far as possible", because, in those days, air-change rates were extremely high (several changes per hour) and many designers assumed that it was virtually impossible to achieve rates as low as, say, 1/4 change per hour. Today we know that rates as low as 0.1 change per hour can be achieved without too much trouble. Today's question is: how low a rate is permissible?

Note concerning latent heat: If, in winter, the air in the house has high humidity, it has much latent heat. When this air is exchanged (in simplest manner) for outdoor air (which has lower absolute humidity), much of this latent heat is lost -- increasing the overall amount of (sensible and latent) heat lost. I am indebted to G.S. Dutt for reminding me of this.

Factors limiting the lowest permissible rate: Most of the factors are obvious. They include:

Build-up of smells; persistence of smells. (Kitchen, bathroom, cigarettes, sweaty bodies, chemicals in hobby room)

Build-up of humidity. (Moisture from cooking, from showers, from the occupants' breathing, from dish washers, from growing plants that are watered daily.)

Build-up of CO_2 from occupants' breathing.

Other factors are:

Combustion products from gas stove. Carbon monoxide. Nitrogen oxides. My guess is that, in superinsulated houses, electric stoves should be used -- not gas stoves.

Combustion products from wood-burning stove (if any), or fireplace (if any). Or leakage of flue gases through gaps (if any) in chimney pipe of furnace (if any).

Build-up of gaseous products, or dusts, from various chemical materials of the house structure or furnishings. In some houses formaldehyde (from certain furnishings, e.g.) can present a problem.

Build-up of radon, a radioactive gas that emits alpha particles. I am told that in houses that have brick walls, or concrete walls or floors, or granite foundations, or employ large quantities of gravel (in the crawl space, or in a thermal storage system) much radon may accumulate if the house is tightly sealed.

The amounts of such materials present in houses of various types is being investigated by research teams at the University of California and elsewhere. Some preliminary results have been presented in Refs. U-520-12 and U-520-13. When and if radon is a threat in a given house, increasing air-change rate to about 0.7 or 1.0 per hour may virtually eliminate the problem.

When are these factors important? In the fall, winter, and spring, because during those seasons houses are kept well sealed, people tend to stay indoors, and much use is made of stoves and heaters. The problems may be especially acute in late fall and early spring: outdoor temperatures are then not very low and accordingly there is little "stack-effect" ventilation. I am indebted to G.S. Dutt for pointing this out to me.

In summer, windows are open and little build-up of harmful gases is to be expected.

Main location of air leakage in woodframe houses: I understand that much of the leakage occurs at:

Sills; cracks above and below sills

Outer edges of doors, where transverse shrinkage may have caused cracks to open up

Fixed frames of windows

Windows proper; especially between parting rails

Holes for pipes; holes in outer walls or roof

Holes for electrical wires

Miscellaneous locations: ducts, registers, thresholds, chimneys, junctions of partition walls and attic, trap door to attic.

Note: I understand that casement-type windows leak much less air than sliding windows leak. Double-hung windows have intermediate leakage rate.

Typical outdoor-vs-indoor pressure difference: I understand that typical differences are in the range from 0 to 5 pascals (0 to 0.02 in. of water). On rare occasions the pressure differences may be as great as 50 pascals.

How does the leakage rate depend on the pressure difference? I have been told that if the cracks in question are very slender, doubling the pressure difference doubles the leakage rate; but if the cracks are wide, doubling increases the leakage rate by only about 40%. A similar conclusion has been advanced by Sherman et al at a recent ASHRAE conference; see A-256; they find the leakage rate to vary between the 1.0 power and 0.5 power of the pressure difference, for very slender cracks and wide cracks respectively.

Experimental method of determining leakage rate: Two methods have been used successfully:

Pressurization method: Employ a blower which forces air into the house (with all windows and doors shut). Measure the overpressure produced within the house, and measure the rate of flow of air delivered by the blower. Compute the flowrate as a function of overpressure. The most commonly chosen overpressure is 50 pascals, corresponding to 0.2 in. of water.

Tracer gas method: Introduce a known quantity of easily detected, inert, gas into the house; diffuse it throughout the rooms. Remeasure the concentration of this gas every few minutes. Find the rate at which it decreases. Equate the rate at which the tracer gas decreases with the rate at which air in the house escapes and is replaced with outdoor air. As tracer gas, ethane may be used. Or sulfur hexafluoride (SF_6).

HUMIDITY

In a typical superinsulated house -- in mid-winter -- the humidity may become excessive (more than about 60% ?) if the rate of air-change is less than about 0.3 changes per hour.

One sign of excess humidity is a tendency for moisture to condense on windows. A small amount of condensate may be ignored; perhaps it will disappear near the start of the next sunny day.

If the windows are triple-glazed, condensation will seldom occur. For it to occur, the outdoor temperature must be very low and the indoor humidity must be very high. The threat is greater if the windows are double-glazed. And if they are (by some miracle of bad design) single-glazed, the amount of condensate may often be enormous.

DEHUMIDIFIERS

One obvious way of reducing high humidity is to use a conventional, portable, electrically powered dehumidifier.

Using such device may be especially good strategy if heat is needed, i.e., in mid-winter. The device not only reduces the humidity but also helps heat the house; it provides heat in two ways (1) the electrical power used provides heat to the room air, and (2) the latent heat of the H_2O is recaptured and contributes to heating the room air.

However, this is not a cheap way to heat a house!

HEAT EXCHANGERS

Two problems can be solved simultaneously by installing a small air-to-air heat exchanger that will continually drive out old air and bring in new air, heat being conserved by transfer of heat from the outgoing air to the incoming air: the incoming air is _fresh_ and its _humidity is low_.

An airflow rate of about 50 to 75 cfm may suffice. Accordingly the heat exchanger can be small and the consumption of electrical power likewise can be small.

Mitsubishi Heat Exchanger

Mitsubishi Co. produces a family of small air-to-air heat exchangers. One of these is the Lossnay Model VL-1500 MC heat exchanger, which uses about 50 watts of electrical power, has an airflow rate of about 65 cfm, and has an efficiency of about 70%. Cost: about $280. Available from Melco Sales, Inc., 3030 E. Victoria St., Compton, CA 90221 or from the New England regional distributor R.F. Walker Assoc., PO Box 271, Newtonville, MA 02160. Such a heat exchanger was installed in January, 1980, in the Leger House in E. Pepperell, Mass. Being compact (about 21 in. x 15 in. x 11 in. in outside dimensions) it can be mounted directly in a portion of a window opening; or may be mounted in an external wall.

The device employs two blowers, which are mounted on a common shaft driven by one electric motor.

The individual air passages in the core of the exchanger are about 1/16 in. in inside diameter, and the flow is laminar. In all, there are one or two thousand such passages. The passages are in two sets: for incoming air and for outgoing air. The flow directions are perpendicular to one another, i.e., at 90°, not 180° (presumably with some sacrifice of efficiency). It may be necessary to occasionally clean the large arrays of tiny entrance-holes, to prevent them from becoming clogged with dust and lint.

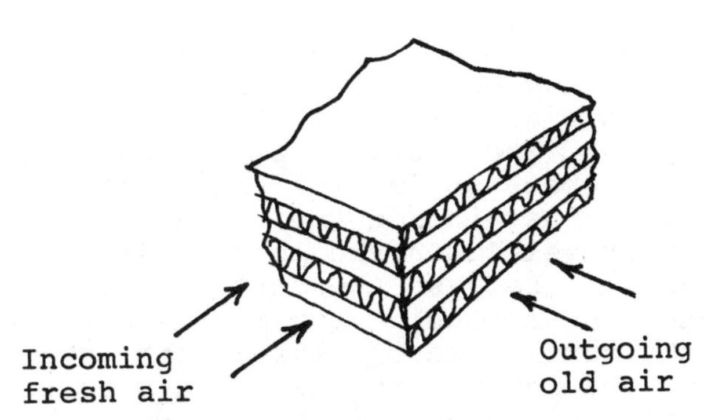

Approximately life-size view of a portion of the Lossnay exchanger

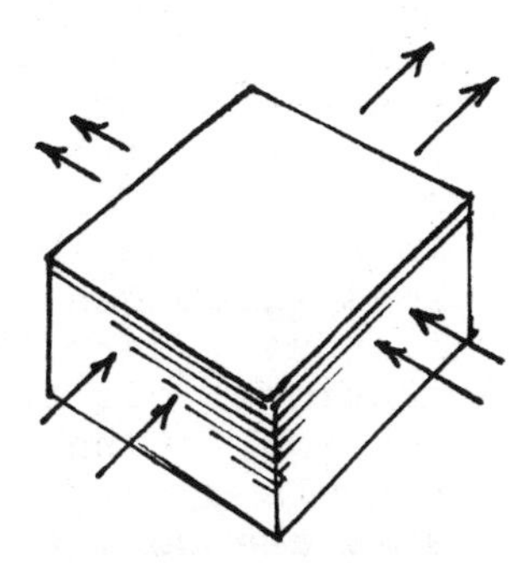

Perspective view of core of Lossnay exchanger

Heat Exchanger Used in Saskatchewan Conservation House

This device, invented by R.S. Besant, R.S. Dumont, and D. Van Ee of the Dept. of Mechanical Engineering of the University of Saskatchewan, and described in detail in a report "An Air to Air Heat Exchanger for Residences" (Bibl. B-251a), appears to be easy to build and performs well. (See figs, next p.)

The exchanger was designed for use in relatively air-tight residences; it is intended mainly for use in winter. Besides exchanging stale air for fresh air, it keeps the humidity in such houses at near-optimum level -- prevents the humidity from becoming too high.

The device is of vertical-parallel-plate, counter-flow type. The heart of the device is a set (stack) of sheets of 0.006-in. polyethylene -- or, more exactly, a single very long sheet that has been folded back and forth many times (with folds rounded to have roughly a ¼-in. radius of curvature) in serpentine manner. There are 36 folds, and 37 flat areas. Between successive flat areas there are ½-in.-thick spacers, or frames, of plywood. Each such frame has two gaps, or ports -- to let air in and out. The 36 airspaces between plastic-sheet flat areas constitute two sets: old-air-out set and fresh-air-in set. The former employs airspaces 1,3,5, etc., and the latter employs airspaces 2,4,6, etc. The assembly as a whole constitutes a rectangular parallelopiped about 65 in. x 24 in. x 19 in. in overall dimensions. It is mounted (just inside the house) with the long axis vertical. The old air travels downward through it, and the fresh air travels upward through it. Because the sets of ports in the plywood frames are well separated, no contamination of one airstream by the other occurs. Because the total area of plastic (the heat-exchange area) is large, because the sheets are thin, and because the thickness of each airspace is great enough to encourage turbulence of flow, the heat exchange is very efficient, being about 90% when the rate of airflow is 100 cfm and about 95% when the airflow is 50 cfm. Flow is maintained by two small (20 w) blowers, one for each of the airstreams.

Defrosting: In mid-winter as much as 25 to 50 lb. of ice may form, each 24-hr. period, in the lower part of the exchanger. Defrosting techniques must be used. For details, see B-251a.

Commercially available device: Early in 1980 a slightly modified device was being produced and sold by Dick Van Ee (member of the Mechanical Engineering Department of the University of Saskatchewan) from this address: RR #3, Saskatoon, Sask. Canada S7K 3J6. The device includes 31 layers with ½-in. spacing, has 340 sq. ft. of heat-transfer surface, is 90 in. long by 24 x 17½ in. in cross section (these dimensions include housing, ductwork, and fans); price is $375 F.O.B. Without ductwork or fans, it costs $200. Power consumption for the pair of fans is 40 watts of 120-v., 60 Hz current. In months when ice build-up is a threat, defrosting is accomplished by using a timer to turn off the fresh-air fan for ½ hour out of every 24 hours.

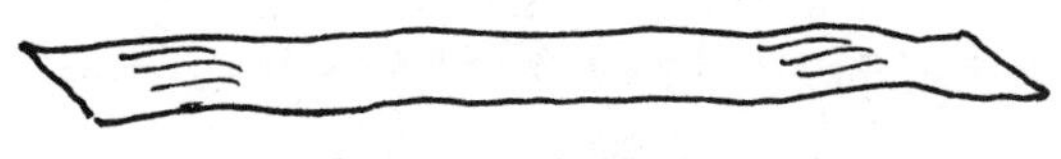

Long sheet of polyethylene

Serpentine-folded sheet

Plywood strip

Gap at end Gap at side

One set of plywood spacers installed, with gaps for passage of old air

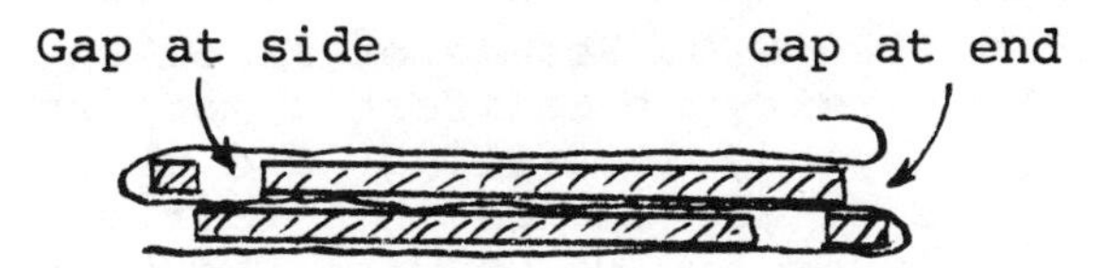

Second set of plywood spacers installed, with gaps for passage of fresh air

Gaps allowing fresh air to exit from exchanger

INDOOR end

Gaps allowing old air to enter exchanger

OUTDOOR end

Gaps allowing fresh air to enter exchanger

Gaps allowing old air to exit from exchanger

Here the assembly is oriented with its long axis vertical. Note separate groups of gaps for separate functions

½-in. plywood spacer

Gap

Gap

Perspective view of one set of plywood spacers and its two gaps

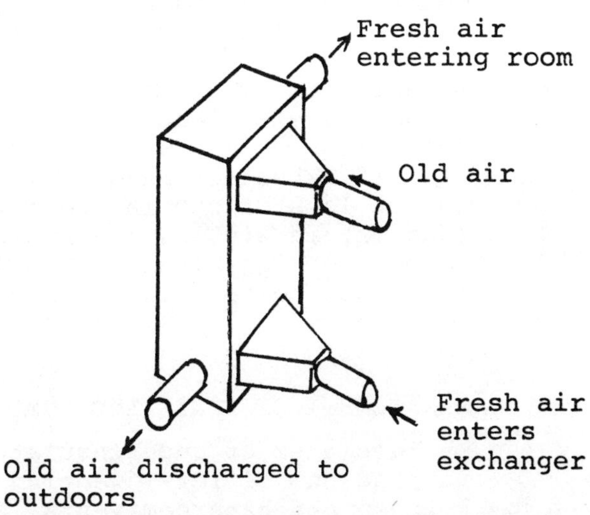

General view of exchanger (in housing) in use

<u>Some design considerations</u>: The airspaces between heat-exchanger sheets should be at least ¼ in. thick in order to permit the airflow to be turbulent, i.e., in order to provide a high-rate of heat-exchange -- high efficiency. In some devices the sheets are highly permeable to water vapor; thus much H_2O can quit the outgoing air, pass through such sheet, and join the incoming air. (In many situations this recovery of the H_2O is highly <u>un</u>desirable: there may be great need to get rid of excessive water vapor.) In cold climates much frost, or ice, may be formed in the heat-exchanger, and accordingly defrosting schedules must be arranged.

Early in 1980, Enercon Industries Ltd. made and marketed an air-to-air heat exchanger, for houses, priced at $795 including fans, controls, and insulated ducting. Late in 1980 a smaller, less expensive exchanger was being developed.

Dollar Saving From Use Of A Heat-exchanger:

In 1980 G.D. Roseme et al of the Lawrence Berkeley Laboratory of the University of California made an analysis of the cost effectiveness of buying, installing, and using a heat-exchanger to increase the rate of air change in a fairly tight house. Their report (U-520-13) includes tables showing the cost effectiveness under a wide range of circumstances. A general conclusion is that the dollar saving is large if the house is in a cold climate, if an expensive kind of auxiliary heat is used, if a low-power blower (in the heat-exchanger) suffices, and if the heat-exchanger has high efficiency. A specific result of importance is presented in the following example.

Example: Consider a house in Minneapolis that has a natural air-change rate of 0.2 per hour. Assume also that auxiliary heat is supplied by an oil-burning furnace, with oil at $1/gal. Suppose now the air-change rate is increased (e.g., by means of a blower) to 0.75 per hour. Much more heat is lost per winter, and much more use is made of the furnace. But suppose that the increase in air-change rate is made by a 75%-efficient heat-exchanger containing a 45-w blower; assume that the cost of buying and installing the exchanger is $700. Then the net saving of heat is great and the saving of money (for oil) is great. The net present value (NPV) of such exchanger is found to be about $2900. Thus the financial benefit is about 4 times the cost. (The NPV is very much less for a house in Atlanta, or if a much-higher-power blower is needed, or if the exchanger's efficiency is much less, or if a much cheaper fuel is available.)

If a heat-exchanger can be designed in such a way that there is no net loss of H_2O in the air, then there is no loss of latent heat. (But if, in winter, the humidity of the house in question happens to be too high, loss of H_2O is desired. To try to retain the H_2O (in order to save latent heat) is unwise.

MANAGEMENT OF VENTS AND HUMIDIFIERS

One designer of superinsulated houses wrote me: "Some people are running humidifiers, venting dryers into the house, not running kitchen and bathroom vents, and complaining of condensation on windows. To compound this problem, many people are turning their thermostats way back or even off at night. Thus a house at high temperature, high humidity, high dew point is cooled to a state at which many windows and doors are well below the dew point. Nighttime cooling of residences can exaggerate humidity problems."

AUXILIARY HEATERS

Some superinsulated houses -- even some in the colder regions of USA -- may get through the winter without need for auxiliary heat. Heat from intrinsic sources and direct-gain passive solar heating may provide 100% of the heat needed.

More often, of the order of 10,000,000 Btu may be required of an auxiliary system. I.e., as much as would be provided by burning about 100 gal. of oil in a 67%-efficient furnace; about $100 worth of oil at Feb. 1980 prices.

The designer could, of course, provide the house with a furnace. But this may be unwise. Furnace plus chimney plus oil tank plus heat-distribution system may cost $3000 to $5000 (?). Operating and maintaining a furnace can be a nuisance. Furnaces can misbehave: produce smoke that might spread throughout the basement, or possibly present a fire hazard. If the electrical power should fail, e.g., in some widespread emergency, the furnace would not run.

Instead, the designer may specify use of a small stove that burns wood or coal. This may work very well if the stove is of well designed type, nicely installed, and capable of being closed off so as to cause no leakage of air when not in use. Preferably it should be supplied with ducted air. But a poorly designed stove, improperly installed, and to airtight, can cause troubles of many kinds.

<u>Making secondary use of gas-type heater for domestic water</u>: This is the scheme used in the Leger House. It seems to me very simple, inexpensive, safe. The gas-type heater (a Paloma, wall-mounted device, approximately the size and shape of a full knapsack) is used straightforwardly to heat the water in a typical, but very well insulated, 40 gal., domestic hot water tank. But when a room thermostat senses abnormally low temperature in the rooms, hot water from this heater is circulated through a short length (about 40 linear ft.) of baseboard radiator. Such supply of auxiliary heat to the rooms is said to be "more than adequate". (For further details on this system, see pertinent section of Chapter 3.)

<u>Making secondary use of an electric-type heater for domestic hot water:</u> Obviously, a somewhat similar scheme could be used if the domestic hot water is heated by electricity. Throughout the greater part of the 24-hour day the electric heating elements are inactive; if one "stole" some heat from this tank to heat the rooms, the heating elements would be powered more often; but the domestic hot water would still be kept as hot as desired. (Should the distribution system be made compatible with keeping the water potable? In that case, no heat exchanger would be needed; the water supplied to the distribution system could be taken directly from the DHW tank.)

Possible combination water-and-air system: Suppose that, in the basement, there is a 40-gal. tank that is kept hot by gas flame or electricity. One could modify (enlarge, expand) the insulating jacket so that, between tank and jacket, there is a 1-in. airspace. One could employ a small blower to drive air within this 1-in.-thick annular space; room air could be circulated (via ducts) to and from this space whenever the room thermostat finds the room temperature to be too low. Admittedly, the heat-exchanger area is small (about 25 sq. ft.); yet it is (I guess) large enough. I expect that the rooms would be kept warm. And the capital investment would be negligible -- assuming the DHW system was already paid for.

Scheme using off-peak electricity and a water tank: Suppose one had, in the basement, an extra-large DHW tank, which was electrically heated. Suppose that the heating elements are controlled so that they are powered only at far-off-peak hours, such as midnight until 5:00 a.m. Then the cost of the electric power would be very small; yet -- because the tank is large and the heat-needs are small -- there would be ample hot water throughout every day and considerable heat could be diverted (if necessary) to heating the rooms (say by means of a within-jacket annular airspace and forced-air circulation system such as is mentioned above).

This scheme exploits the very long carrythrough (very slow cool-down) of a superinsulated house. To keep such a house warm, it is not necessary to produce the heat at the same time that the heat is needed; it can be produced shortly after midnight, and used, for example, on the following evening. Such "produce-now, use-later" option is highly attractive.

The scheme has this added feature: if the electrical power is off during a 16-hour emergency, the house can still be kept warm (or at least fairly warm) thanks to the heat that is stored in the large DHW tank. Some heat can be extracted from it by gravity convection. Or the entire jacket could be temporarily removed, to speed the natural extraction of heat from the tank. (A small bank of photovoltaic cells would suffice to run the very small blower, which might be of 40-W power rating.)

Note: off-peak electric power may or may not be especially cheap, depending on what kind of fuel the electric plant uses (nuclear? oil?) and whether or not peak demand is so great that construction of a new billion-dollar plant may be necessary.

Scheme using regular refrigerator as a heat-pump: It has occurred to me that an ordinary refrigerator, in use to keep food cool, could be made to serve also as a heat-pump (small capacity heat-pump). Run two tiny ducts from 50°F basement to the refrigerator, and arrange for a small amount of basement air to circulate through the interior of the refrigerator. Then the refrigerator motor will run harder, and will deliver (to kitchen) an additional amount of heat, via the coil at the back of the refrigerator. Whether this idea has any real merit, I do not know.

"Santa Claus" scheme using heat stored in massive basement walls etc. and upgraded by a high-COP air-conditioner: See Appendix 3.

Can the designer omit a heat-distribution system? Perhaps he can. If the auxiliary heat is discharged at some central location in the house, it may diffuse through the house fast enough without any formal distribution system. E.H. Leger's experience has shown this to be true. Thus much money may be saved.

RETROFIT SUPERINSULATION

Is it feasible to superinsulate a typical existing house? I don't mean merely "increase the heat-saving considerably". I mean increase it enormously. How would one go about this? Is it feasible?

The subject is a big and important one. Unfortunately I have failed to find much information on it, other than the information presented in the following paragrpahs.

Walls: W.L. Shick has proposed (in an unpublished report of 1976) greatly improving the insulation of the north, east, and west walls -- but perhaps not the south walls inasmuch as their heat-loss is partly compensated by the solar radiation incident there. His proposal, if I understand it correctly, is this: Indoors, on a north, east, or west wall, install an additional set of 2x4 studs, 24 in. apart on centers; place this set 2 in. from the existing wall, so as to leave a 2-in. airspace. Then fill the 2-in. airspace and also the spaces between studs with fiberglass; that is, install 5½ in. of fiberglass. Then install a vapor barrier, seal it well, and install drywall. Install additional glazing in the windows; extend the window sills etc. Perhaps the original window-frame trim can be saved and then reinstalled (at a location about 6 in. farther toward the interior of the room).

Window sash-weight boxes: These may be filled with insulation and then sealed (caulked). Unfortunately, filling them with insulation immobilizes the sash weights; but one can still open the windows -- by lifting more strongly.

A scheme that has been used successfully by Stewart Coffin of Lincoln, Mass., is to install, along the interior box-face closest to outdoors, a ½-in. sheet of Thermax -- which barely touches the sashweights and still allows them to move up and down freely. Also, all pertinent cracks are caulked.

It will be enormously helpful to our country if practical ways of superinsulating existing buildings can be developed.

IS IT A GOOD IDEA TO COMBINE SUPERINSULATION AND A SMALL ACTIVE-TYPE SOLAR COLLECTOR?

I have heard this view expressed:

> "if your house is superinsulated, use of a small active-type solar collector is especially effective. The collector can be small and inexpensive. The combination of superinsulation and a small active-type collector is likely to be especially successful."

But I have heard a superinsulation advocate deny this. He is convinced that, if an active-type collector is needed, the house has not been designed correctly. He believes that a well-designed superinsulated house simply does not need an active collector.

I am sympathetic to this view. Also I see this additional argument against the small collector: a small collector may have especially high cost per square foot. Even if it is very small, it may have to have the same sensors and controls as a large collector, and perhaps the same inspection and maintenance requirements. One might argue: "If all you need is a very small collector, let's have none at all!".

A slightly different view is this: If you wish to take in more solar energy, make the passive system larger. Increase the area slightly; or install some outdoor reflectors that will increase the amount of solar radiation received by the passive system. Also, employ thermal shutters on cold nights.

WHAT IF THE OWNER DEMANDS A LARGER AREA OF WINDOWS?

If the owner, desiring a larger view, demands that the window area be increased somewhat, what can the designer do? He can accede to this demand, preferably specifying windows that are wide but not very high: if they are only 3 or 4 ft. high, they can be very effectively shaded by the eaves in summer -- thus the danger of overheating the rooms in summer can be avoided. Also, he may do well to call for:

thermal shutters or shades for all large windows, for reducing heat-loss on cold nights,

increased thermal mass in the south and west rooms (perhaps all rooms). An additional (second) layer of gypsum boards may be used on the walls. Floors and/or ceilings can be made more massive. Concrete block partitions may be used.

Chapter 6

SUPERINSULATED HOUSES: PERFORMANCE

PITFALLS ASSOCIATED WITH THE TERMS: WINTER, AVERAGE OUTDOOR TEMPERATURE IN WINTER, DEGREE-DAY VALUE

I am indebted to G.S. Dutt for pointing out the pitfalls associated with these innocent-looking terms.

Winter: Smith and Jones live in the same town in New England. Smith, who lives in an uninsulated house, says: "Winter extends from Sept. 1 to May 31. Throughout that period I make much use of my furnace." Jones, who lives in a superbly insulated house, says: "Winter extends from Dec. 5 to March 15. I use my furnace in that period only." To these two men, living in the same town, the term winter has very different meanings.

Average outdoor temperature in winter: Smith, who uses his furnace from Sept. 1 to May 31, says: "The average outdoor temperature in winter is 45F." Jones, who uses his furnace only from Dec. 5 to March 15, says: "The average outdoor temperature in winter is 33F." To these two men, the term average temperature in winter has very different meanings (different values).

Such differences in meaning can play havoc with attempts to compare the performances of houses said to be situated in places having the "same length of winter" or "same average outdoor temperature in winter". Equal havoc may result from naive attempts to "correct" for differences in "length of winter" or in "average outdoor temperature in winter".

Degree-day value: The men who (long ago) defined degree-day and adopted 65F as the base temperature have gotten heating engineers into a jam. All might have gone well if the decision had been made to use 70F as base (and to assume the normal room temperature to be 70F); for, then, a degree-day value for any location would indicate straightforwardly the temperature deficiency, or shortfall in outdoor temperature, relative to the desired indoor temperature.

Unfortunately, those men decided to throw in allowances for the typical amounts of intrinsic heating and direct-gain passive solar heating -- assuming a typical amount of insulation in walls, ceilings, etc., and typical amounts of air leakage. At that time the inclusion of such allowances made fairly good sense inasmuch as most houses were more or less alike in the respects in question. But today, houses differ widely in these respects; there are wide differences in amounts of insulation, amounts of air leakage, and amounts of south-facing windows. Whereas 65F may be a good choice of base temperature for Smith's house, which has no insulation and few electrically powered devices, 45F may be an appropriate value for Jones' house, which is superbly insulated and includes a dozen kinds of electrically powered devices.

Such considerations can play havoc with attempts to compare the performances of houses that are said to be situated in places having the same degree-day value.

LIMITATIONS TO THE "DEGREE-DAY" CONCEPT

As far as computations of gross heat loss from a superinsulated house are involved, the usual degree-day concept -- with its base temperature of 65°F--is fairly useful. The heat-loss is very nearly proportional, for a given house in a given location, to the degree-day value for that location. If the degree-day value pertinent to a certain house were to increase (somehow!), the gross heat loss would increase correspondingly.

But as far as computations of amount of heat needed (from the sun, via south windows, and from the furnace), it is of little use. This is because, for a superinsulated house, the standard base temperature 65°F is highly inappropriate. Even when the outdoor temperature is as low as 45°F (say), such a house may remain warm (about 70°F) with no solar-energy intake and no use of the furnace -- because the insulation is so superb and the intrinsic heat sources are so large.

For such a house, and for use in computing the necessary amounts of heat from sun or furnace, a base temperature of about 45°F may be appropriate.

Other Bases For Degree-Day Data

Can one easily convert a table of monthly degree-day base-65°F values to some other base? No. Not easily. It does not suffice merely to make a simple subtraction.

A rough conversion can be made by assuming that the smoothed curve of daily outdoor temperature as a function of time throughout the year is a sinusoidal curve. If one knows (a) the average value for the year and (b) the amplitude of the sinusoidal component, one can make a good approximation to any other desired temperature. (I am indebted to N.B. Saunders for pointing this out to me.)

To obtain a fully accurate table of DD values using a different base temperature, one should go back to the original data: the day-by-day values, and make the appropriate subtraction from each. Recently, a group at the National Oceanographic and Atmospheric Administration, National Climate Center, Federal Bldg., Asheville, NC 28801, has compiled degree tables using a great variety of base temperatures from 25°F to 75°F. (I am indebted to William Braham for telling me about these tables.)

Way back in 1965 Warren Harris (H-62h) warned of the shortcomings of degree-day-base-65F data and proposed empirical correction schemes.

In an excellent article published in 1979, R.H. Bushnell has described a masterful solution to the "degree-day base temperature" problem. He uses temperature data with respect to absolute zero (or other very low starting value); he prepares graphs showing (for each location) how many days have each choice of temperature ("2 days have ave. temp. 5°F, 5 days have ave. temp. 6°F...); then -- after tabulating the data -- one can quickly compute the full-winter degree-day values pertinent to any choice (or many choices) of base temperature (such as 65°F, or 45°F etc.) For details, see Bibl. B650.

HEAT-LOSS IS NOT PROPORTIONAL TO (Δ T)

Often a designer assumes, for simplicity, that the gross heat-loss from a building is directly proportional to (Δ T), the difference between indoor temperature (70F, say) and outdoor temperature. Having made this assumption, he finds it reasonable to employ a standard, winter-as-a-whole degree-day figure, especially if he feels he can make a reasonably correct estimate of the pertinent base temperature (e.g., 65F, or 55F, or whatever) and can obtain the degree-day figure pertinent to this base.

Although roughly correct under various circumstances, such procedure sometimes leads to very wrong results, especially if the building is in a windy location, is several stories high, is very leaky, and has a large basement. Why does it give wrong results? Because:

It takes no account of changes in windspeed. The greater the speed, the greater the rate of in-leak of cold air and the rate of out-leak of warm air. I.e., the greater the rate of heat-loss.

It takes no account of stack effect: the tendency (especially in a tall building) of cold air to leak into the lower part of the building and warm air to leak out of the upper part -- even when there is no wind. The volume of in-leaking air increases with (Δ T) and the heat-loss-per-unit-volume likewise increases with (Δ T); accordingly the stack-effect heat-loss is roughly proportional to the square of (Δ T).

It takes no account of the fact that the basement may, typically, be much cooler than the rooms. On a day when the rooms are losing a moderate amount of heat, the basement may be losing none (because it is at the same temperature as the outdoor air).

It takes no account of the fact that, in the first part of the winter, the basement (if cool) receives much heat from the underlying earth. The temperature of the earth a few feet below the basement is highest in the fall (not in mid-summer!) and lowest in the spring (not in mid-winter!). The basement-earth temperature cycle lags weeks or months behind the cycle of outdoor temperature.

Recognizing these complications, an engineer may use a more elaborate calculation: he may estimate the gross heat-loss of the building by adding four terms:

term proportional to (Δ T)
term proportional to windspeed*
term proportional to $(\Delta T)^2$
special term intended to correct for lag in basement-earth temperature.

How useful is the degree-day figure here? It is useful in connection with the first term -- term linear in (Δ T). But it is not directly applicable to the other terms.

(I am indebted to N.B. Saunders and R.S. Dumont for explaining to me some of the above-described complications.)

*If the house in question is in an exposed windy location, heat-loss due to wind-caused air-leakage may be ten times greater than heat-loss due to stack-effect air-leakage. I am indebted to N.B. Saunders for this information.

CRITERIA RECENTLY USED IN JUDGING THE PERFORMANCE OF A PASSIVELY SOLAR HEATED HOUSE

In recent years, persons judging the performance of a passively solar heated house have used some of the following measures or criteria:

A. Total amount of useful heat provided by the solar heating system

B. Percent of heat-need provided by the solar heating system

C. Reduction in amount of auxiliary heat needed

D. Reduction in cost of fuel consumed

E. Solar savings fraction (SSF)

F. Amount of auxiliary heat needed

How valid, or pertinent, are these criteria (to a typical, passively solar heated house)? None is fully valid!

Criterion A is poor inasmuch as the solar heating system may create a large fraction of the heat-need. Adding 200 sq. ft. of single-glazed south windows greatly increases the amount of solar radiation captured, but it also greatly increases the heat-loss on a cold night. In some extreme cases the increase in heat-need can exceed the increase in useful amount of solar energy collected! The amount of auxiliary heat needed may be increased!

Criterion B is poor for the same general reason: even if an added area of glazing increases the percent of heat-need provided by the sun, it could be that the amount of auxiliary heat needed is also increased!

Criterion C is fairly good -- up to a point. But it entirely overlooks the important question: "Is a furnace still needed?" Saving an additional $100 or so on oil is fine, but nowhere so near so fine as eliminating the need for a $4000 furnace system.

Criterion D has the same limitation: it takes no account of whether the furnace can be eliminated.

Criterion E, routinely used by the group at Los Alamos Scientific Laboratory, is doubly defective: (1) It takes no account of whether the furnace can be eliminated, i.e., shuts its eyes to a possible saving of, say, $400 to $4000, and (2) the method of choosing the comparison house (prerequisite to computing the solar savings fraction) is vague and artificial. It is vague in that one is supposed to (mentally) replace the "solar aperture" with a panel that blocks all flow of energy, i.e., transmits no solar radiation or other radiation and has no thermal conductance. (What, exactly, is a solar aperture? South windows? Southeast windows? East windows? North windows?) It is artificial in that the resulting structure, which in some cases is totally dark and provides no view, is unlivable and therefore cannot properly be called a house. Worse: this structure consists of a mismatched combination of features: features specially chosen as parts of a solar house and features squarely inappropriate (windows covered by energy-blocking panels). I see little purpose in comparing an actual solar house with an artificial structure that is unlivable and includes a mismatched combination of solar and anti-solar features.

Criterion F is relevant but not fully relevant. It too takes no notice of savings from having a smaller auxiliary heating system. Such savings sometimes exceed the savings on fuel.

SOME BETTER CRITERIA

Some criteria that appeal to me (as regards performance of a typical passively solar heated house) include:

How small is the cost of the auxiliary heating system needed? For example, if only a single portable electric heater is needed, the cost may be $25. If a 70,000 Btu/hr oil furnace (with oil tank, chimney, heat distribution system, etc.) is needed, the cost may be $4000.

How small is the amount of auxiliary heat needed per year? If the amount is only 2,000,000 Btu, or will cost only about $25 per year, the owner will be very pleased.

How uniform will room temperature be throughout the daytime and evenings?

Will water-filled pipes be safe with respect to freeze-up threat even if, in February, supplies of oil, gas, and electricity fail?

Will the house tend to remain cool in summer? Or will it be very hard to keep cool? Some passive solar houses are disasters in summer.

Of course, durability, attractiveness, etc., also deserve much attention.

Convenience, too, deserves much attention. A house with a huge area of south windows may require a huge area of thermal shutters or shades, and operating these may be a nuisance (and if the occupants are away, these devices may not be operated at all).

Any significant deprivations, too, should be taken into account. A passively solar heated house that is largely underground may provide no view to east, north, or west, and, in summer, may permit no through-draft from west to east (the direction of the prevailing wind). Also it may provide very little privacy -- if all rooms face a south terrace to which tradesmen, visitors, etc., have access.

Attention should be given also to freshness of air and to maintaining a comfortable humidity.

If a designer insists of comparing the thermal performance of a passively solar heated house with that of some other house, let this other house be selected for the excellence of its thermal performance. If the best thermal performance is provided by a superinsulated house, make the comparison against the best superinsulated house of corresponding size.

Fourth day thermal vulnerability: Some passively solar heated houses require enormous amounts of auxiliary heat after several cold sunless days in February, while others may require little or no auxiliary heat at such times. I have sometimes used the expression "fourth day thermal vulnerability" to mean the amount of auxiliary heat needed (to keep the house at 70F) in the fourth day of a many-day period in February that is sunless and has typical February temperature. Only if this "vulnerability" is very small can the designer contemplate omitting the furnace.

PERFORMANCE CRITERIA PERTINENT TO SUPERINSULATED HOUSES

What performance criteria should be used in judging the behavior of a superinsulated house -- if it has no furnace, requires practically no auxiliary heat? I.e., if practically all of the heat needed is supplied by human bodies, light bulbs, cooking stove, etc., and by solar energy received via moderate-sized view-windows? What can be said about solar heating if there is no explicit solar heating system: no added window area and no added thermal storage?

Perhaps these are the best criteria:

- Cost and durability of the house as a whole
- Extent to which the indoor temperature stays close to 70F
- Extent to which the humidity stays close to 45%
- Extent to which the air stays fresh
- General livability of the house, with attention paid to possible bonuses such as (a) view in all four directions (N,E,S,W), (b) wide range of view from each corner room, (c) attached greenhouse (?). Note: a greenhouse can produce vegetables, flowers, fragrance, but it can also produce bugs, moisture, and continual responsibility.

Comments on these criteria:

- Cost of the house is a reasonably definite concept. It can be estimated prior to construction and evaluated fairly accurately after construction.
- Durability may be hard to estimate until the house has been in use many years.
- Temperature-close-to-70F is a dubious goal. Some persons prefer other temperatures, and some prefer to have some variation in temperature: changes are invigorating, stimulating. But in any event, how can one assign a number to comfort? How can one mold it into a quantitative measure? I do not know. (I am indebted to G.S. Dutt for explaining the difficulty to me.)
- Extent to which the humidity stays close to 45%: here again, tastes differ and there is no obvious way to quantify the resulting comfort.
- Extent to which the air stays fresh; again, tastes differ, and freshness may be hard to define in a general way, and hard to measure.
- General livability: hard to define, hard to quantify.

This much seems certain: persons who like to use computers to compare the performances of solar houses are in for a rude shock when they try to apply their computers to superinsulated houses. Most of the interesting performance parameters of superinsulated houses are outside the domain of technology: they are psychological. (Psychology is not an exact science. Some cynics say that it is not a science at all! If you ask a psychologist to recite some of the main laws of psychology, he will stare at you in dumb amazement.)

HEAT-NEED FROM FURNACE PER (SQ. FT. OF FLOOR AREA, WINTER DEGREE-DAY VALUE)

Here I present some values of the amount of auxiliary heat needed (e.g., from furnace) per unit area of the floorspace of heated portion of house, per unit of wintertime degree-day value (relative to base 65F) of the site. (I use this abbreviated name for the parameter: Performance parameter #10, or PP-10.) The unit of this parameter is Btu/(ft^2 DD).

Note that the gross heat-need is much greater; but much of this is satisfied by intrinsic heat sources and by the sun. PP-10 pertains just to the shortfall: the amount the auxiliary heating system must supply if the rooms are to be kept at 70F.

The logic behind the parameter PP-10 is highly defective. I discuss the defects in a later paragraph.

House	PP-10 parameter (Btu/ft^2 DD) (Warning: values may be rough estimates only, and the logic underlying the parameter is faulty)
Leger House, prior to stopping last leaks and with no regular occupancy; four coldest months only	2.5
Same, but for entire winter. (Note: no additional auxiliary heat was needed in the additional months included.)	1.3
Same, but assuming normal occupancy and elimination of last airleaks	0.9
Same, but with addition of thermal shades	0.4 to 0.6 (guess)*
Same, but regarding the (warm) basement as part of the living area	0.2 to 0.3 (guess)*
Saskatchewan Conservation House, with thermal shutters in regular use	0.6
Kirkwood House (predicted value)	1.0
"Typical house" as specified by Mid-American Solar Energy Complex	12.5
"Baseline moderately insulated house" as specified by D. Lewis and W. Fuller in Solar Age Dec. 1979, p. 31.	7.5
Goal adopted by Mid-American Solar Energy Complex: "Solar 80 Criterion".	2.5

*The reason that I make casual guesses here is that the values are, in any event, of almost no importance, as long as they are below, say, 1. The annual cost of auxiliary heat is less than the cost (at a good restaurant) of dinner for four persons.

Defective Logic Underlying the PP-10 Parameter

The above-mentioned parameter, the unit of which is Btu/(ft^2 DD), is defective in many ways. The parameter is used as a criterion in judging the thermal performance of a given design of house; designers regard a value of 15 as poor, 5 or 10 as fair, and 1 to 3 as excellent. In fact, the parameter is a poor criterion of thermal design of house.

It is a poor criterion because the values depend not only on the design of the house but also on: (1) the thermal climate, (2) the amount of solar radiation, (3) local shading by buildings and trees, (4) prevalence of wind, (5) manner in which the house is used.

Thermal climate: A superinsulated house may have a PP-10 value of zero if the house is in a DD-3000 location and a value of (say) 2 if it is in a DD-8000 location. The trouble is that the amount of auxiliary heat needed by a superinsulated house is very far from being proportional to the DD value if the (usual) base 65F is used. (Use a more appropriate base? This won't work: you would need different bases for houses insulated to different extents.) Also, even in locations that have the same DD value (e.g., 4000), a given house may have very different PP-10 values depending on how widely the temperature varies in winter; if the range is very narrow, PP-10 might be 1, but if the range is large, PP-10 might be 2.

Amount of solar radiation: A superinsulated house with very small window area may have almost the same PP-10 value whether it is in a very sunny region or a very cloudy region; but a house with 500 sq. ft. of single-glazed south windows will perform very much better in a sunny region than a cloudy region; even assuming that the regions have the same DD value, the PP-10 values might be as different as 2 (for the sunny region) and 10 (for the cloudy region).

Local shading by buildings and trees: A given house may have high or low PP-10 depending on whether the area to the south is clear or contains many tall buildings and trees.

Prevalence of wind: PP-10 may be high or low depending on the prevalence of high winds. Especially if the house has no caulking, weatherstripping, or vapor barrier.

Manner in which the house is used: To obtain a remarkably low PP-10 value, the owner of the house could crowd in more occupants, reduce the air-change rate from 0.5 to 0.25 changes per hour, encourage the taking of frequent showers and frequent use of cooking stove, increased use of electric lights and appliances. Also, if the basement is warm, he could install a sofa there and include the basement area as part of the living area, which (if the house has only one story) doubles the living area and reduces PP-10 by a factor of 2. (Note: None of these ploys are cheating: all are legitimate. But they greatly impair the comparability of PP-10 values estimated by different people for different houses.)

In summary, the PP-10 parameter depends heavily on the thermal, solar, and wind climate, and on local shading, and depends heavily also on how the house is used and whether the basement is regarded as part of the living area. Because the parameter depends on so many things (of which building design is only one), it is a poor parameter to use in judging building designs.

Some of the above-listed drawbacks to the parameter are not serious if the houses being compared are poorly insulated and sealed and the PP-10 values are greater than, say, 10. But especially when the houses are well insulated and sealed, and have PP-10 values below 5, the drawbacks loom large. Two houses rated 1 and 3 might both be rated 2 (say) if moved to similar sites in the city and occupied by families having similar life-styles.

I am amazed that I have <u>nowhere</u> seen warnings by others as to the faulty logic behind the parameter in question. They seem to accept the parameter as truly meaningful and of outstanding value in comparing different houses. (Note added in proof: R.S. Dumont et al, in a May 1980 paper "Measured Energy Consumption of a Group of Low-Energy Houses", include a warning that for a "low-energy house" the parameter is not linear with respect to degree-day value"; also they express the need for designating a "reference family" that generates a fixed amount of intrinsic heat.)

A Comment On The Solar 80 Criterion

Notice that the criterion refers to a difference: the difference between (1) <u>gross</u> heat need and (2) heat supplied by intrinsic sources and by the sun. To satisfy the criterion, the designer can choose either (a) to keep the gross heat need very small, i.e., provide very effective insulation and keep the air-change rate very low, or (b) to keep the amount of intrinsic heat and/or amount of solar energy received very large. Or he can do a little of both.

The important point is that the designer may decide to use a large area of passive collection of solar energy. In an extreme case, he might call for, say, 500 sq. ft. of south-facing windows. Although satisfying the Solar 80 criterion, such house would <u>not</u> conform to my definition of a superinsulated house. (My definition stresses keeping the gross heat need very small. Also it stresses keeping the "fourth day thermal vulnerability" small and keeping the summertime cooling load small.)

GROSS HEAT-LOSS

My impression is that, in a cold (7500 DD) location, a 1500-ft^2 superinsulated house has a gross heat-loss of about 40,000,000 Btu. See Chapter 3 sections on Lo-Cal House and Saskatchewan Conservation House.

Presumably the value is roughly proportional to the winter degree-day value. (This is _not_ true of the amount of heat which the furnace must supply.)

SOLAR FRACTION

My impression is that, in a cold (7500 DD) location, a 1500-ft^2 superinsulated house receives of the order of 1/3 of its gross heat input from solar energy. See Chapter 3 re Lo-Cal House.

In an extremely warm location, all of the heat needed may be supplied by intrinsic sources. Solar heating may then play a minor, or trivial, role -- or, on the whole, may do more harm (from overheating) than good.

SOLAR SAVINGS FRACTION

Even for passively solar heated houses of ordinary type, this term, much used by solar heating experts at the Los Alamos Scientific Laboratory, is of dubious significance, little value. It relates a given passive solar house to a fictitious structure (of similar size and shape) in which all the solar apertures (windows etc.) have been replaced by idealized panels that have no transmittance, no conductance. A detailed list of shortcomings of "Solar Savings Fraction" is presented on a previous page.

CARRYTHROUGH

My impression is that a typical superinsulated house has very long carrythrough. The 1/e time constant may be in the range from 50 to 120 hours. In a cold climate, on a cold mid-winter night with no intrinsic heating, room temperature may drop only 0.7 to 1.0 F degrees per hour. Normally there _are_ intrinsic heat sources and the rate of cool-down is lower, and when the sun comes out the rooms begin to warm up again.

INCREMENTAL COST OF SUPERINSULATION

In the Saskatchewan Conservation House the incremental cost (relative to the cost of a house of comparable size that has ordinary-1978-quality insulation) is said to be about $4000 minus $1000 = $3000 (see B-250, B-251c). The cost of added insulation and improved vapor barrier is $4000, and the saving from using a smaller heat distribution system is $1000.

In the Leger House the incremental cost is less: surely as small as $1000 and perhaps actually zero. Here the main savings were from having no furnace and only a very small heat-distribution system (only 40 linear feet of baseboard radiation - and it turns out that even a smaller amount would have sufficed).

My impression is that, ordinarily, the incremental cost of superinsulation is only a very few thousand dollars, and sometimes less. "Sometimes less" will be more frequently applicable as builders become more familiar with the new techniques and the special precautions.

In comparing the costs and benefits, one should not overlook the "fringe benefits" such as lack of drafts, a more satisfactory humidity level, greater exclusion of highway noise, reduced fire risk (from having no furnace), greater security with respect to failures of oil supply, electrical supply. Also, keeping the house cool in summer is easier and cheaper.

<u>What about the cost of the space preempted by the additional insulation?</u> One friend said: "If all four walls of a house are 5 in. thicker than a typical 1975 wall, the useful floor area is decreased by about 60 sq. ft. If the house costs $50 per sq. ft., this loss of floor area represents a loss of 60 x $50 = $3000. This is very large!" But another friend pointed out that if one slightly enlarges the house, so that there is no decrease in floor area, the cost of the enlargement is far less than $50 per sq. ft. -- inasmuch as no additional doors, partitions, bathrooms, windows, electrical outlet boxes, etc., are needed. He estimates the total incremental cost to be about $1200, made up equally of materials and labor. (From Robert Corbett of National Center for Appropriate Technology I have learned of this idea: Apply the added insulation (expanded polystyrene, say) to the <u>outer</u> faces of the walls, and let this insulation extend out beyond the vertical planes defined by the foundations, i.e., let it "hang over". Then there is no preemption of indoor space, and yet there is no need to enlarge the foundations. Protect the added external insulation with, say, Drivit.)

PERFORMANCE IN SUMMER

In summer, superinsulated houses keep relatively cool -- if the windows and outside doors are kept closed during the day and are opened wide during the cool parts of the night. The thick insulation of the walls, roof, etc., keeps out heat during the day. The wide eaves shade the south windows. The east and west windows are not shaded, but they are quite small. (Equip them with awnings?)

In cooling a house at night, it may be worthwhile to employ an exhaust fan to drive out the old air and draw in new. In an article in Alternative Sources of Energy, #41, Jan.-Feb. 1980, S. Baer explained the desirability of maintaining a large flow of incoming cool air at night -- so that the rate at which the walls typically give off heat to the room air will be matched by the rate at which the room-air heat can be driven outdoors.

The Leger House, in the summer of 1979, required use of an air conditioner for only a very few days (more exactly, few hours).

HOW EFFECTIVE ARE EAVES IN EXCLUDING RADIATION IN SUMMER?

They are much less effective than one might assume. Typically, they cut the amount of solar radiation entering the south windows by only about 50%, according to an article by Utzinger and Klein in Solar Age, 23, 369 (1979).

Why is this? Because diffuse radiation and ground-reflected radiation are of considerable importance and are scarcely reduced at all by the eaves. Also, if the house aims more than 10 or 20 degrees away from straight south, much direct radiation enters the windows at about 8:00 or 9:00 a.m. and 3:00 or 4:00 p.m. Also, in September and October the sun is so low in the sky even at noon that eaves help only a little; the rooms may become too hot.

The moral? Provide awnings or shutters. These can do a far better job of excluding radiation. Note: If awnings or shutters are used during hot summer days and thermal shutters or shades are used during cold winter nights, a very different area of windows, and a different distribution, may be preferable. See Chapter 5.

DIFFERENT LIFE-STYLES CAN AFFECT PERFORMANCE

Obviously, the thermal performance of a house is highly dependent on the habits, or life-styles, of the occupants. If they leave doors and windows open, or open and close the outer doors dozens of times a day, much auxiliary heat may be needed. If they cook cabbage and smoke cigars (or take frequent showers or keep kettles of water boiling for long periods), much faster air-change may be needed to prevent build-up of odors or high humidity.

Especially if the house uses very little auxiliary heat, i.e., especially if the house is superinsulated, small changes in life-styles may make large percentage changes in amount of auxiliary heat used.

Whenever a given house seems to be performing much better or much worse than expected, take a close look at the occupants' life-styles!

CRY OF DESPAIR FROM CONNECTICUT

A friend in Connecticut takes a dim view of superinsulation. In effect, he says:

> The discovery of how very effective superinsulation can be is very depressing. Almost any building, irrespective of size, shape, etc., can be superinsulated. Thus the challenge to architects is destroyed. Instead of happily finding ingenious ways to put solar energy to work, and instead of finding how to strengthen the warming bond between sun and man, architects can plod ahead with the dull, outmoded designs of yesterday, merely bolstering them up with superinsulation.

I do not subscribe to this view. I do not rank challenge to architects above keeping buildings warm. Preserving challenges is not a true goal. Keeping buildings warm is. Anyway, plenty of challenges remain.

GENERAL ENTHUSIASM

In talking with designers, builders, and occupants of superinsulated houses, I find great and uniform enthusaism. There is a strong consensus that the houses keep warm, that very little auxiliary heat is needed, that the extra costs involved (for added insulation, added care, etc.) are small, and that the whole approach is highly cost-effective.

Some wonder whether the added costs are as small as claimed; they suspect the true cost is larger than indicated, especially if the contractor is not familiar with superinsulation.

Others raise questions as to the quality of the air. They are uneasy about air-change rates of only 1/4 or 1/2 per hour.

Some are categorically down on houses that have no greenhouse, no very large sunny area.

Commenting on superinsulated houses in general, John K. Holton, architect and energy conservationist, has said: "This is potentially an approach that is extraordinarily practical. What it means is that one can design window areas based on the appropriate needs of the internal spaces and be constained neither to minimize window areas or slavishly locate them all on the south exposure. The ability to cut loss/load way down allows one to tailor windows, and their gain, to the satisfaction of the design. This is true design freedom - with responsibility."

A recent report by an international group lends strong support to the view that superinsulation is especially cost-effective. See Bibl. item R-260. See also p. 12.

Chapter 7
DOUBLE-ENVELOPE HOUSES:
PRELIMINARY CONSIDERATIONS

INTRODUCTION

Definition: "Double-envelope house" is defined in Chapter 1. Central to the concept is the greenhouse on the south side and the path for gravity-convective flow of air.

Non-new aspects: The idea of having an integral greenhouse is very old. The idea of encouraging gravity-convective flow of air is also very old. The idea of transferring heat to a large thermal mass by flow of hot air is very old. The idea of including much insulation in walls is very old.

New aspects: What is new, I understand, is the idea of (a) dividing the north wall into two portions -- inner and outer -- with a 1-ft. airspace between, (b) similarly dividing the roof into two portions with an airspace between, (c) providing a sub-floor space of comparable dimensions, (d) incorporating a large greenhouse in the south part of the house, and (e) connecting all of these to form a complete loop-like path for gravity convective flow of air.

I understand that Lee Porter Butler was the main pioneer of this new approach and that Thomas Smith was the first to complete and occupy a house employing this approach. I have been told that Malcolm Lillywhite played an important role in the early thinking, and that several others contributed. Recently, many other designers and builders have extended or modified the early designs.

In 1978 Tom Smith and L.P. Butler published a book describing and promoting double-envelope houses. Long before this, Butler had evolved a number of house designs of double-envelope type. In the autumn of 1977 Smith completed his double-envelope house at Olympic Village (near Lake Tahoe, CA) and he and his family moved in. Soon many articles on this house were published.

In 1980 Donald Booth wrote a book on double-envelope house built by him or others in, or near, New Hampshire. See Bibl. B-415.

DEVELOPMENTS IN NORWAY

In November 1976 Johs. Gunnarshaug and others of the University of Trondheim and the Norwegian Institute of Technology, Trondheim, Norway, described a building design employing a wall system that has a thin (few centimeters) airspace, or plenum, between an outer insulated wall and an inner insulated wall. In the next few years the idea was extended to include thicker airspaces, and there was increasing emphasis on gravity-convective (thermosiphon) circulation in these airspaces. Experimental houses with such double-envelope walls on roofs, and with integral greenhouses on the south sides, were built and analyzed. Results were said to be excellent. For further details, see Chapter 11.

WILL THE IMPORTANCE OF THE HISTORY OF THE DOUBLE-ENVELOPE PRINCIPLE INCREASE?

Today there is uncertainty as to the merits of various features of the double-envelope houses such as are being built in America and in Norway. Some features may be very helpful, others less helpful. Design changes may be made. Until it is clear which features are truly successful, questions as to who invented the features, and when, may be of only moderate interest.

The interest will be much greater when and if performance studies show, for example, that:

Much heat flows around the convective loop,

Much heat is actually deposited in the beneath-crawl-space thermal mass,

Bathing the inner walls of the living space with circulating warm or luke-warm air is significantly helpful,

Transporting heat from greenhouse to crawl-space thermal mass merely by gravity convection is as reliable and cost-effective as using a duct and blower,

The amount of auxiliary heat needed is as small as, or nearly as small as, the amount needed in a superinsulated house,

The cost of the house is in reasonable relationship to the cost of a superinsulated house.

I have made no systematic effort to explore the history of the double-envelope idea. I sometimes wonder whether other designers, years ago, toyed with using a double-envelope design and, rightly or wrongly, decided the idea was not promising.

PRINCIPLE OF OPERATION

The principle of operation of a double-envelope house is discussed in detail in Chapter 10.

Here is a brief summary of the principle.

On a sunny day:

> The greenhouse, incorporated in the south side of the house, receives much solar radiation. The air in the greenhouse becomes very hot, and rises, and flows northward along the within-roof space, flows downward in the within-north-wall space, flows southward in the crawl-space, then -- to complete the cycle -- flows upward into the greenhouse via grilles or slots in the greenhouse floor.
>
> This flow of air carries heat from greenhouse to the crawl-space, and some of the heat enters the crawl-space earth and is stored there.
>
> En route from greenhouse to crawl-space, the warm air bathes the inner roof and inner north wall, contributing to comfort in the rooms contiguous to that roof or wall.
>
> The flow of hot air from the greenhouse, and the return flow of cool air from the crawl-space, helps keep the greenhouse from becoming excessively hot.
>
> The greenhouse performs other functions: it may contribute warm and pleasantly humid and fragrant air to the rooms, it may serve as a pleasant place to sit or work, and vegetables and flowers may grow there.

On a cold night:

> The greenhouse becomes colder than the crawl-space earth, and a reverse-flow of air then begins, carrying heat from that earth to the greenhouse -- to keep it from becoming extremely cold. Doors and windows between south rooms and greenhouses are kept closed so that those rooms will remain warm even if the greenhouse is fairly cold.

Why does hot air from the greenhouse flow all the way around the convective loop during a sunny day? Does not hot air want to go up, not down (i.e., not down into the crawl-space)? The answer is that the hot air from the greenhouse cools somewhat in traveling along in the within-roof airspace, and cools more as it travels downward within the north wall airspace; accordingly the air in this latter space is much more dense than the air in the (very hot) greenhouse, and thus gravity pulls harder on the former than on the latter, causing the desired loop-type flow.

If the air is cooled as it travels (within roof and north wall) toward the crawl-space, how can it impart heat to the crawl-space earth? The answer is that it cools only to an intermediate extent: it is still fairly warm when it reaches the crawl space. In summary, it cools enough to produce the desired speed of flow, but not so much as to preclude the desired delivery of heat.

Chapter 8

DOUBLE-ENVELOPE HOUSES: FOUR HISTORIC EXAMPLES

INTRODUCTION

Here I present detailed, systematic descriptions of some of the more important and better known double-envelope houses. The descriptions are arranged alphabetically by state.

I understand that the Lake Tahoe House (Smith House) was the first to be completed and occupied. It is certainly the best known.

Of the four houses described here, I have visited only two: Burns House and Mastin House.

CALIFORNIA

Olympic Village, California (on north shore of Lake Tahoe, 140 mi. NE of San Francisco)

Lake Tahoe House (Smith House): PO Box 2356, Olympic Village, CA 95730. 39°N. Altitude: 7000 ft. An 8150-DD location. Completed in Dec. 1977.

Architect of initial design: Lee Porter Butler. Some design revisions were made by Thomas Smith.

Owner, occupant: Thomas Smith. Funding: private.

Building: Two-story building with 1800 sq. ft. of living area, or 2250 sq. ft. including the integral greenhouse. Plan dimensions: 33 ft. (E-W) x 30 ft., or 33 ft. x 38 ft. including greenhouse. The building faces 20 deg. E of S. In winter trees shade the greenhouse after about 2:30 p.m.

The first story includes living-dining-kitchen area, bedroom, bathroom, and (on west side) air-lock entrance. The second story includes a bedroom and two other rooms and a bathroom. Beneath the first story there is a 3-ft. crawl space. There is no attic.

The total window area on the E and W sides is small and double glazing is used. There are no windows on the north side. Total south-facing glazing area (greenhouse and clerestory windows): 390 sq. ft.

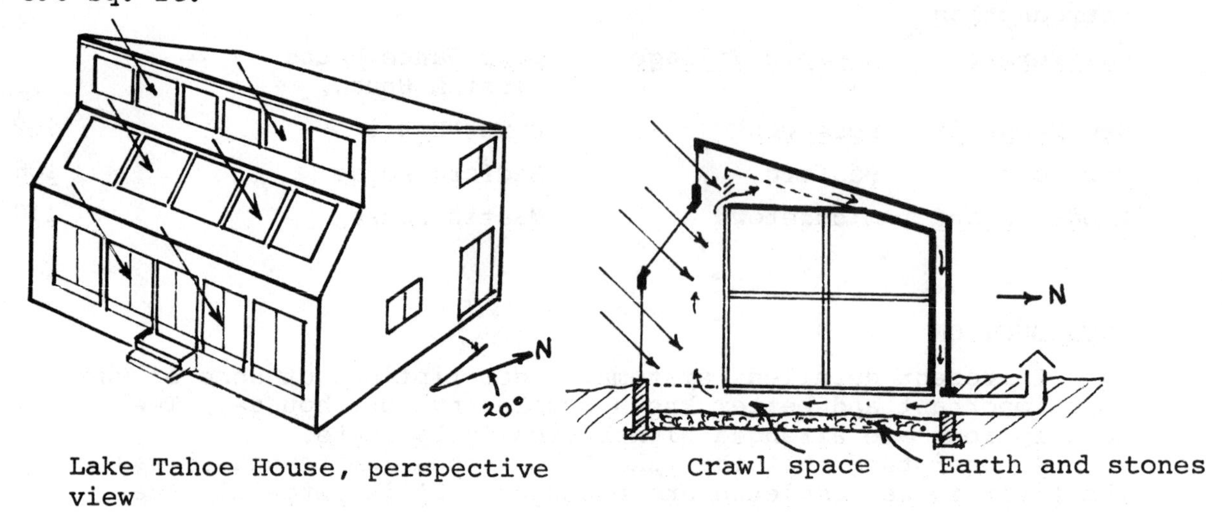

Lake Tahoe House, perspective view

Crawl space Earth and stones

Greenhouse: The 250-sq.-ft. (33 ft. x 8 ft.) two-story integral greenhouse has a total glazed area of 330 sq. ft. About 53% of this is vertical and 47% slopes 60 deg. from the horizontal. Glazing is double. The greenhouse floor is largely unobstructed and is slotted; the slots are 1/4 in. wide and 6 in. apart on centers. Total open area: 10 sq. ft. A small fraction of the greenhouse floor is covered by earth-filled containers in which plants are growing. A large fraction of the wall between greenhouse and adjacent rooms is glazed; there are double-glazed sliding glass doors and single-glazed windows. There are no thermal shutters or shades. Above the greenhouse there is a row of 6 vertical south-facing clerestory windows with a total area of 60 sq. ft. Some of these windows are openable.

Double-envelope system: This includes greenhouse, north wall system, roof system, and crawl space system.

Roof system: This consists of an inner roof and outer roof separated by a 12-in. airspace. Each roof is well insulated. On the underside of the outer roof there is a 0.004-in. polyethylene vapor barrier.

North Wall system: This includes an inner wall and outer wall with a 12-in. airspace between. These two walls contain 3½ and 5½ in. of fiberglass respectively, and provide R-11 and R-19. The airspace is largely clear, constituting a plenum 15 ft. high and 31 ft. in E-W dimension. On the inner face of the outer wall there is a polyethylene vapor barrier.

Crawl space: Extending under the entire area of the living space and the greenhouse there is a 3-ft.-high crawl space. Below it is a 2-ft. layer of earth and small stones resting on a 1¼-in.-thick layer of urethane foam. Enclosing the crawl space region there is an 8-in.-thick concrete foundation wall insulated on the outside.

In summary, hot air from the greenhouse is free to circulate, by gravity convection, into the within-roof airspace, thence downward in the within-northwall airspace and into the crawl space, and thence to the greenhouse via its slotted floor. In passing along the crawl space, the moderately hot air imparts some heat to the earth and to the vertical walls of the crawl space.

On cold nights a reverse circulation occurs. This slightly warms the air in the within-roof airspace and the within-north-wall airspace and thus helps reduce heat-loss from the pertinent rooms -- increasing the comfort there. Also, heat is imparted to the greenhouse; such heat prevents the greenhouse interior from becoming colder than about 45F on extremely cold nights and colder than about 55F on moderate nights.

Auxiliary heat: This is provided by a wood-burning stove in the living room. An electric space heating system was installed to satisfy code requirements, but has not been used.

Cooling in summer: Hot air from the greenhouse is vented via several of the clerestory windows. This process draws in air via a large-diameter, long pipe buried deep in the earth outside the house. The incoming air is cooled as it passes along the pipe, which is surrounded by cool earth.

Cost: About $54,000, not including some labor provided by owner.

Performance data: By December 1979 little detailed numerical data on air temperatures, drop in air temperature around the airflow circuit, flow rates, or temperature changes in the crawl-space earth was available. However, the important fact is that the occupants have found the house to be very comfortable in winter and in summer. The owner estimates that one cord of wood has been consumed in the stove each winter.

References: See book by Tom Smith and L.P. Butler (Bibl. item S-290). Also *Popular Science* for Dec., 1979, p. 54 - 58, *Omni* for Nov. 1979, p. 50, *New Shelter* for Sept. 1980, pages 83 - 85.

NEW HAMPSHIRE

Boscawen, New Hampshire (12 mi. NNW of Concord)

Burns House: 43½°N. A 7500 DD location. Completed Sept. 1979.

Architect and builder: Community Builders (Don Booth et al), Canterbury, NH.

Owner and occupant: Warren and Mary Burns

Cost of house (not including land): $75,000

Performance studies: being made by Community Builders with assistance by N.B. Saunders, C. Seavers, R.O. Smith with funding by DOE.

Building: One-story, 2-bedroom, double-envelope-type, woodframe house with integral greenhouse, partial (22 ft. x 15 ft.) basement under SE bedroom, 2-ft. space under the rest of the floored area of the house, and, attached to west end of house, a 2-car garage.

Total enclosed first-story area of house: 1280 sq. ft. (40 ft. E-W, 32 ft. N-S).

Total area of floor of within-inner-envelope space: about 1150 sq. ft.

Percent of volume of house sacrificed to upper, north, and lower airspaces serving the convective loop: about 30%

Percent of house volume that is within the inner envelope: about 46%

Warning: All sketches, plans, etc., are approximate only: not precise and not to scale.

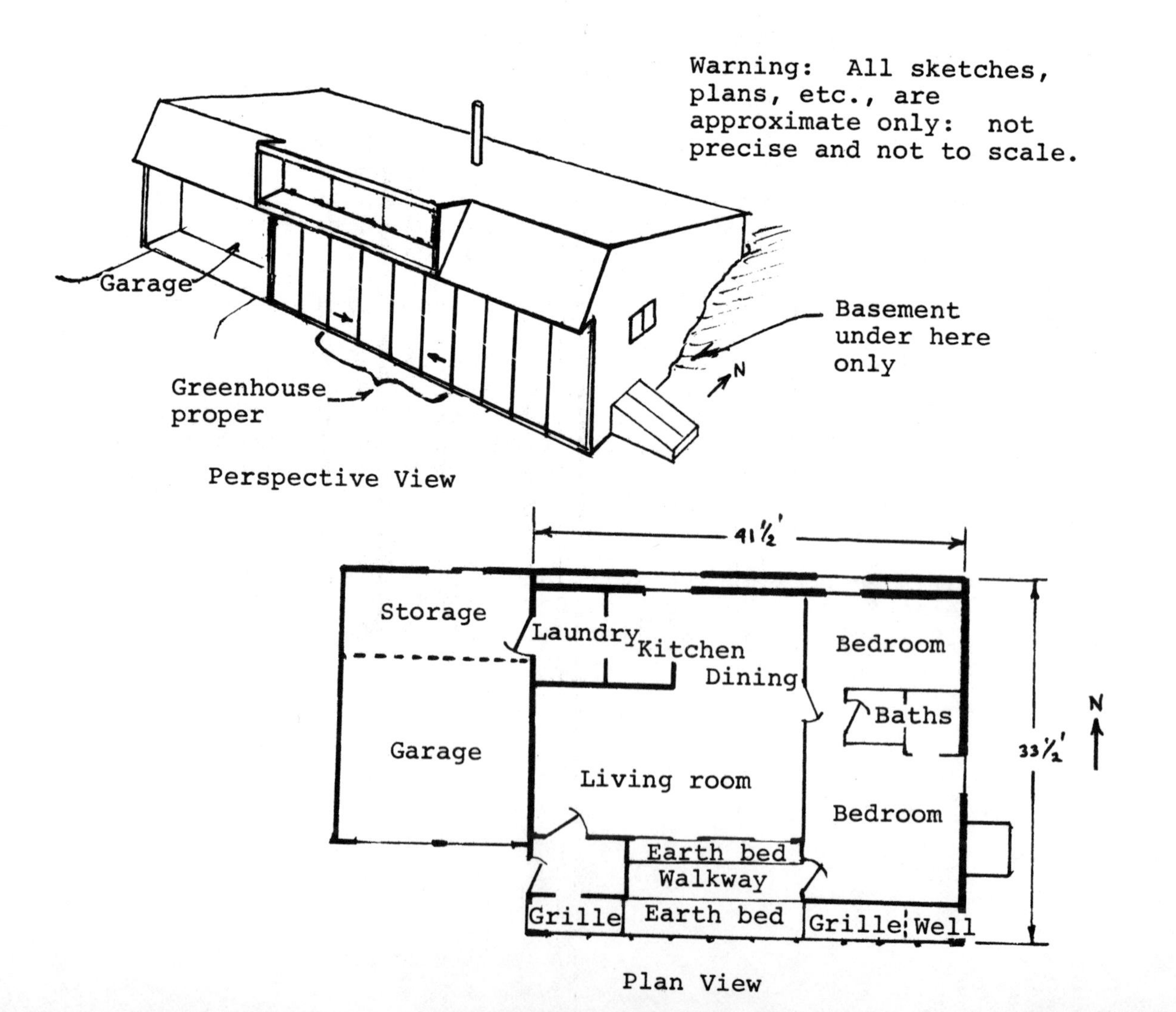

Perspective View

Plan View

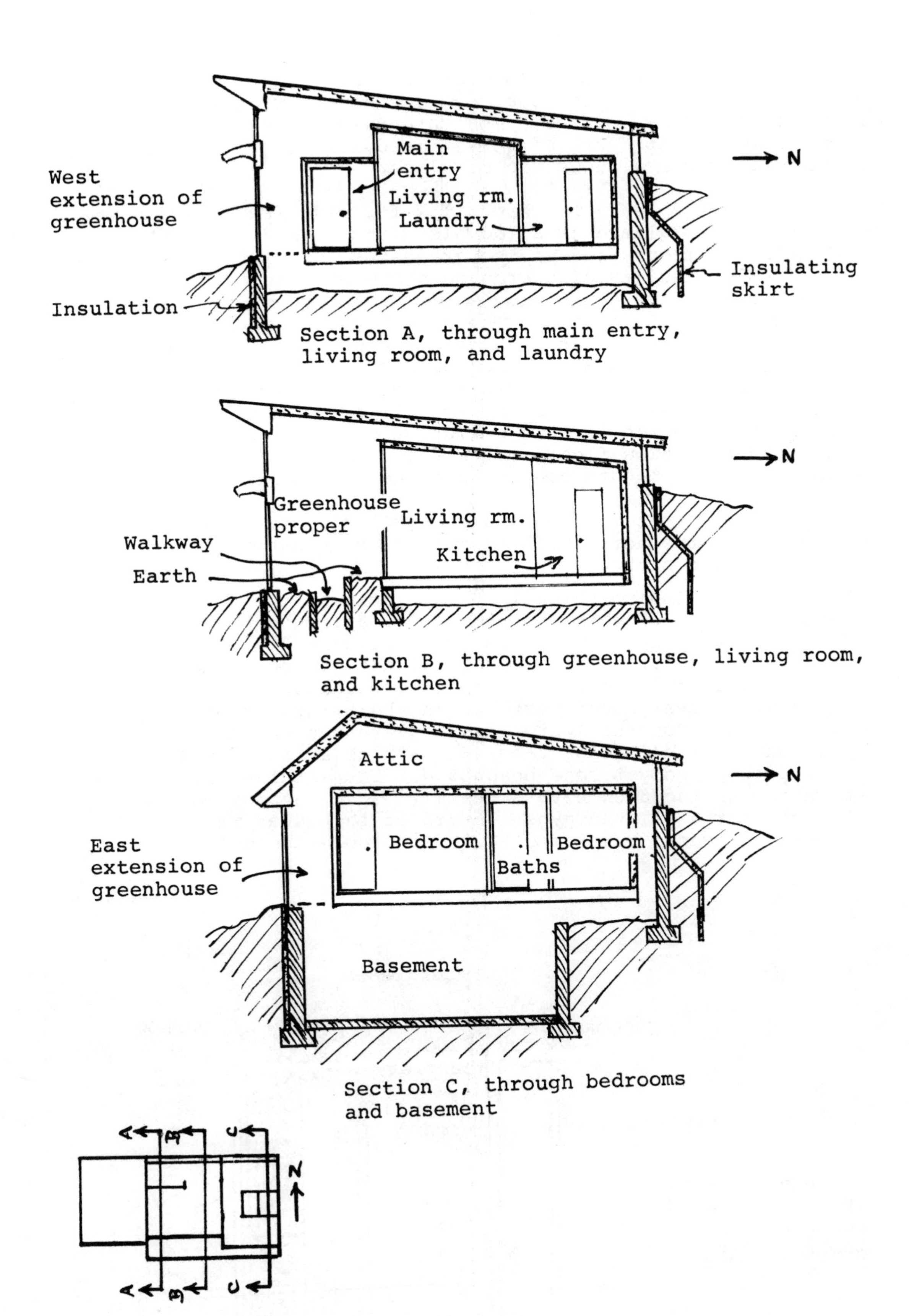

Section A, through main entry, living room, and laundry

Section B, through greenhouse, living room, and kitchen

Section C, through bedrooms and basement

Key to cross sections presented above

Windows: Total area: 388 sq. ft., of which 345 sq. ft. is on the south side, 8 sq. ft. is on the east, and 35 sq. ft. is on the north. All of these windows are double glazed. Likewise the windows between greenhouse, or greenhouse extensions, and the rooms are double glazed. The three windows at the top of the greenhouse proper are of acrylic and are hinged at the bottom. All other windows are of glass.

Greenhouse: This is 13 ft. high, 15½ ft. wide (E-W), and 8 ft. deep (N-S). The glazed area (double-glazed) is 200 sq. ft. About half the floor area of the greenhouse consists of earth beds and half consists of a crushed stone walkway. Most of the area between the greenhouse and adjacent rooms is glazed (double-glazed) with glass). The greenhouse space is open at west and east; that is, at each side there is an extension that is 2½ ft. deep (N-S). Here too much solar radiation is received. The total glazed area of greenhouse and its extensions is 300 sq. ft. Air can flow into the greenhouse via these spaces -- which in turn communicate (via gratings or wells) with the below-floor airspaces. Two of the 4 outer panels of the greenhouse can be slid open in summer.

Double-envelope system: The double-envelope, or convective loop, includes the north wall, crawl space, basement, greenhouse, roof, and attic.

North wall: The north wall system includes two walls (inner and outer) with an 8-to-10-in. airspace between. The inner wall contains 3½-in. of fiberglass (R-11). The uppermost part of the outer wall is of woodframe construction and includes 7 in. of fiberglass (R-22); the lower above-ground part and the below-ground part are of 8-in. concrete which is insulated on the outside with an outward-flaring 2-to-4-in.-thick skirt of moisture-proofed polystyrene beadboard. The within-wall airspace is faced on both sides with ½-in. gypsum board. Behind (north of) the gypsum board of the outer wall there is a vapor barrier of 0.006-in. polyethylene.

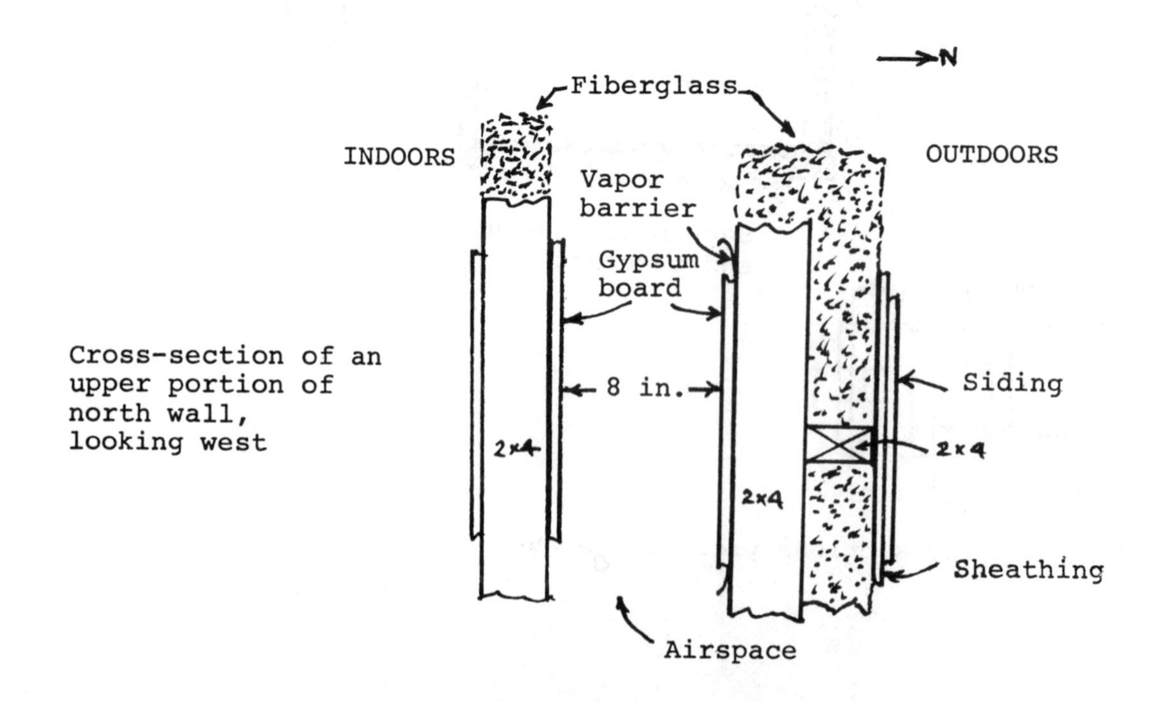

Cross-section of an upper portion of north wall, looking west

Crawl space: Under the entry, kitchen, living room, and NE bedroom there is a 30-inch-high crawl space. Air is free to flow here.

Basement: Under the south bedroom there is a 22 ft. x 15 ft. basement with 8-in. concrete walls. Floor: 4 in. concrete slab resting on undisturbed earth.

Roof: The roof system above most of the rooms includes an inner and outer roof, with 12-in. airspace between. Each of these roofs contains 6 in. of fiberglass (R-22); total thermal resistance of fiberglass in roof is R-44. On the underside of the outer roof there is a vapor barrier of 0.006-in. polyethylene.

Attic: Above the ceilings of the bedrooms there is a small attic, providing an airspace 5½ ft. high at maximum and about 1 ft. high near the minimum (at the north). The attic floor has 6 in. of insulation and the attic ceiling likewise has 6 in. of insulation.

Air in the space below the entry can flow upward, via a 7 ft. x 2 ft. grating, into the west extension of the greenhouse. Air from the space below the bedrooms can flow upward, via an open region (well), into the east extension of the greenhouse. The grating mentioned consists of 1/8-in. x 1 in. steel bars mounted (on edge) 1 in. apart on centers; thus about 87% of the grating area is open. In the well that is mentioned there is a 16-ft.-high ladder.

Air in the greenhouse extensions is free to flow laterally into the greenhouse proper, whence it can flow upward into the within-roof airspace or attic airspace.

In summary, hot air from the greenhouse and the other south spaces is free to circulate in a complete loop by gravity convection. It can circulate into the sub-roof space, then northward, then downward in the north wall airspace, into the sub-floor airspace, then southward, and thence upward to return to the south spaces and greenhouse.

Vapor barriers: As indicated above, these are of 0.006-in. Polyethylene and are located on the inner face of outer roof and inner face of outer north wall. There are vapor barriers also on the east and west walls. There is no vapor barrier in the floor or crawl space.

Auxiliary heat: This is supplied by a wood-burning stove in north central part of the living room.

Domestic hot water: Final heating is of conventional electric type. The water is solar-preheated in a slender vertical tank (enclosed in transparent plastic) in east extension of the greenhouse.

Cooling in summer: Three large vents (acrylic-glazed) at top of greenhouse may be opened manually. Various windows and sliding glass doors may be opened.

Performance data: Little data was available as of 2/13/80. In the first half of the 1979-1980 winter only a small fraction of a cord of wood (about 1/20 cord) was burned in the above-mentioned stove.

References: 1980 book by Donald Booth; see Bibl. item B-415. New Shelter for Sept. 1980, pages 74 - 77.

NEW YORK

Bedford, New York (40 mi. NNE of NY City)

Bedford House: In Bedford Township. Mailing address: 16 Glenwood Lane, Katonah, NY. Jan. 1980. 41°N. A 5500 DD location.

Architect: Lee Porter Butler. Assoc. architect: Frank Santillo.

Builder, owner, occupant: Joshua Arnow. House built for sale.

Building: Two-story, 3-bedroom, 3-bathroom, woodframe, double-envelope house with integral 36-ft.-x-10-ft. greenhouse. There is a full basement and detached 2-car garage (with loft) and breezeway. Plan-view outside dimensions of house proper: 42 ft. x 36 ft. First story includes living room, study, dining area, kitchen, bathroom, vestibule, and greenhouse area. Second story includes three bedrooms and two bathrooms. Main entrance is of air-lock (vestibule) type.

Area of living space kept warm 24 hours a day in winter: 2380 sq. ft.

Percent of volume (of house) kept warm 24 hours a day: about 37%

Percent not kept warm 24 hours a day (greenhouse, basement, spaces above second story, space in north wall): about 63%

Windows: Total area: 564 sq. ft. About 400 sq. ft. (of this total) pertains to the greenhouse. Areas on west, north, and east walls are 42, 80, and 42 sq. ft. All windows are double-glazed; Thermopane is used on south and north, and storm windows are used on west and east.

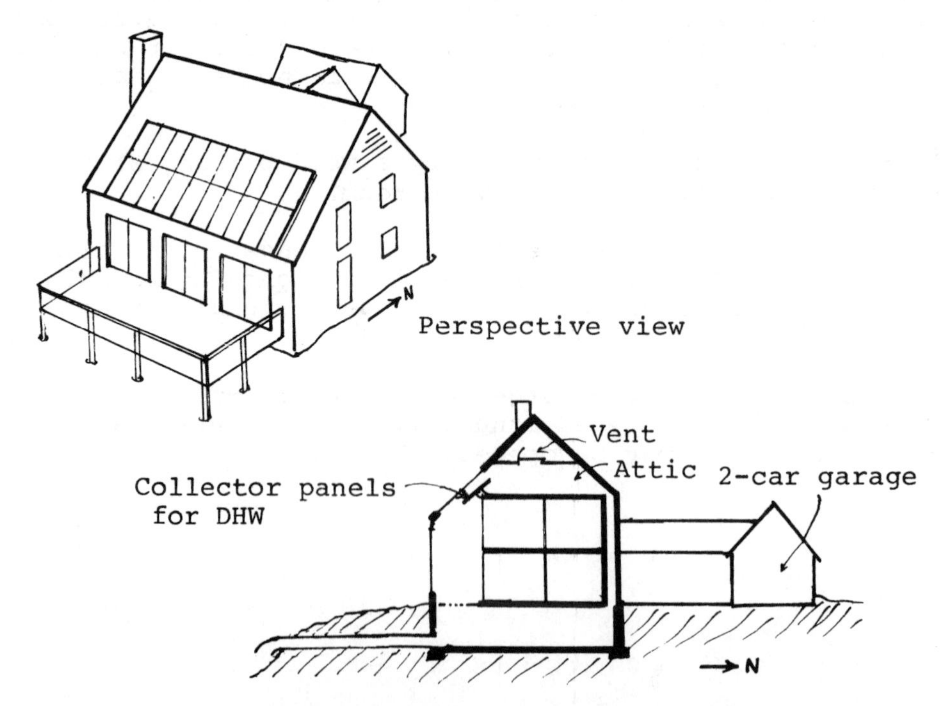

Perspective view

Vertical cross section, looking west

Greenhouse: This is 36 ft. x 10 ft. in plan dimensions and is 16 ft. high. Floor area: 360 sq. ft. About 70% of the glazed area slopes 45° and the rest is vertical. The glazing is Thermopane. The greenhouse floor consists of tiles. There are two long slender grilles in the floor: each is 36 ft. x 1 ft. and is of metal. Combined area of grilles: 72 sq. ft. The grilles permit flow of air from basement to greenhouse or vice versa. No thermal shutters are used.

East wall: (likewise west wall) This includes a single row of studs: 2x4 studs, with 3½ in. fiberglass insulation between. Outer face of wall is covered with 1 in. of High-R Sheathing (essentially identical to Thermax) which has aluminum foil on both faces. Face of wall that is toward room includes a vapor barrier.

Double-envelope system: The double-envelope system, or convective loop, includes the north wall, basement, greenhouse, and below-attic space.

North wall: The north wall includes two walls (inner and outer) with a 9-in. airspace between. The inner wall employs 2x4 studs 24 in. apart on centers, and 3½ in. of fiberglass (R-11). A 0.004-in. polyethylene vapor barrier, installed on warm side, is covered by ½-in. gypsum board. The outer wall employs 2x4 studs 16 in. apart on centers and includes 3½ in. of fiberglass. The vapor barrier on the warm side has no covering. On the outer side of the outer wall there is plywood sheathing, tar-paper, and shingles. The 9-in. airspace, or plenum, in the north wall system is 36 ft. in E-W dimension and two stories high.

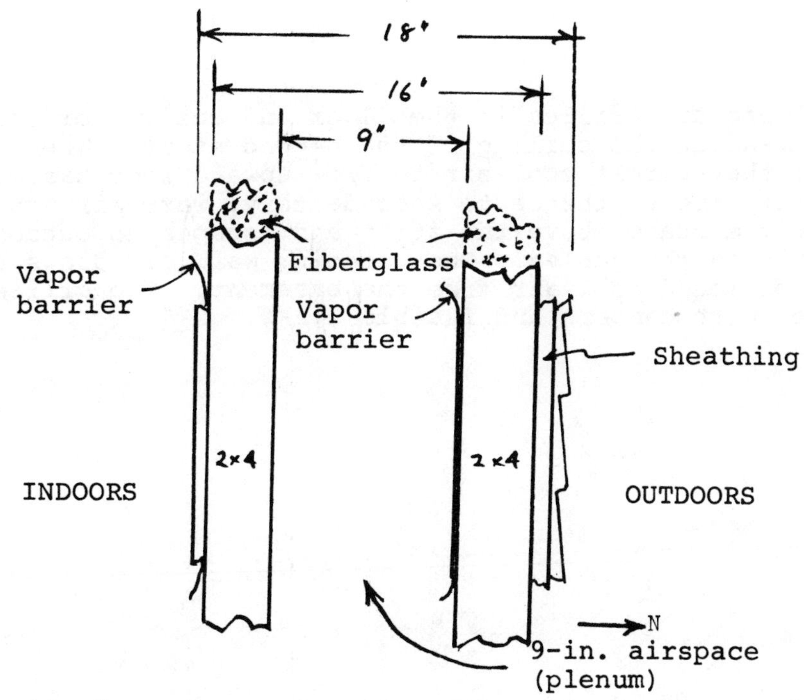

Vertical cross section of north wall, looking west

Basement: The full basement has a 4-in. slab floor, with no insulation, no vapor barrier. The 9½-ft.-high foundation walls are of 12-in.-thick concrete blocks with a 1½-in. layer of Styrofoam insulation on the outdoor side. There is also a Styrofoam skirt that extends outward about 4 ft.

Greenhouse: This has been described above.

Below-attic space: This space is between the attic floor and the second-story ceiling.

Attic: This is about 8 ft. high. On the floor of the attic there is a 6-in. layer of fiberglass, and on the underside of it there is a 0.004-in. polyethylene vapor barrier.

Walls between south rooms and greenhouse: These consist in large part of glass: Thermopane.

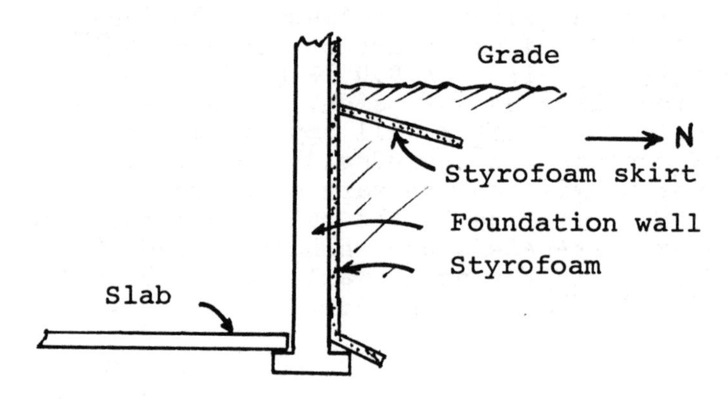

Vertical cross section, looking west, of north foundation wall.

Grilles: There are grilles in the floor and ceiling of the first story and in the ceiling of the second story. When open (in summer) they permit cool air to flow upward from basement to first-story rooms, thence to second-story; warm air travels upward into the space above the attic and escapes to outdoors via the vents in the gables. The outgoing warm air draws up the above-mentioned cool air from the basement. The grilles are equipped with dampers and fusible links.

Vents: In the attic ceiling there are vents -- foam-core, steel-clad doors (6x3 ft.) laid horizontal (when closed) or tilted upward by means of ropes and pulleys. There are also vents at the two gables; each of these is triangular -- 10 ft. wide at the base and 5 ft. high at the center; each is left open in summer.

Fire dampers: Above the airspace in the north wall there is a linear set of four dampers, or flaps, of sheetrock which close automatically (e.g., to stop spread of fire) when fusible links are exposed to very high temperature. There is another set of dampers high up in the south side of the house, a few feet above and to the north of the DHW collector panels.

Vapor barriers: These are of 0.004-in. polyethylene. The locations are indicated above.

Humidity: No information is yet available as to whether any humidity problems or condensation problems are to be expected.

Auxiliary heat: There is a Heatilator fireplace with ducted outdoor air. The room walls include wiring to accommodate electric base-board heaters or wall heaters. Each bathroom includes a small electric heater. There is no furnace, or stove.

Domestic hot water: 60% of the heat needed is provided by an active water-type collector situated in uppermost part of greenhouse, just beneath the Thermopane glazing. Collector area: 100 sq. ft. Absorber plates are of copper. Hot water flows by thermosiphon to insulated tank in attic. Back-up heat is by electrical heating elements in the tank.

Cooling in summer: Hot air in the house can flow upward via vents between space below attic and attic itself and via the vents in the east and west gables. The grilles in floors and ceiling of rooms help also. As warm air leaves the building, outside air may be drawn in, by thermosiphon, via two pipes, each 2 ft. in diameter and 60 ft. long, buried 6 ft. deep in the earth on the south side of the house. Each end of each pipe is screened, and at the indoor end there is a damper, or close-off valve. Deciduous trees shade a large fraction of the south side of the house in summer. There is no air conditioner.

Cost: Total cost of building, including materials and labor (but not land, or architect's fee, or interest): about $135,000. I.e., about $48/sq. ft.

Performance data: Not yet available.

References: N.Y. Times, Sunday, Oct. 14, 1979. Home Section. Also personal communications from J. Arnow.

RHODE ISLAND

Middletown, Rhode Island
(3 mi. NE of Newport)

Mastin House: 1355 Green End. Ave., Middletown, RI 02840. 41½°N. A 6000-DD location. (401) 849-4200. Occupied in March 1979, completed December 1979.

Architect: Lee Porter Butler of Ekose'a.

Builder: Medeiros Bros. Construction; also Robert Mastin.

Owner and occupant: Robert Mastin. Privately funded.

Cost (not including land): $82,000. Performance studies: by Brookhaven National Laboratory (Ralph Jones et al) with cooperation of SERI and the University of California at Berkeley.

Building: Three-story double-envelope building with 2100 sq. ft. of living area, or 2600 sq. ft. including greenhouse and space below it. Plan dimensions: 33 ft. (E-W) x 22 ft., or 33 ft. x 30 ft. including greenhouse. There are four bedrooms and two bathrooms. Also a small attic, shallow sub-basement space, and, at east end of house, an attached 1-car garage. Bldg. faces exactly south.

The lowest story (basement, mainly below grade) includes a family room, bedroom, bathroom, and utility room. The basement floor is of wood and has R-11 insulation. The main story contains kitchen, dining, living areas and, at north, a vestibule-type entry. The top story includes two bedrooms, a bathroom, and a small balcony projecting into the greenhouse space. The sub-basement space is 12 in. high. The attic space (above top story space) is 12 ft. wide at the base and 6 ft. high at the center.

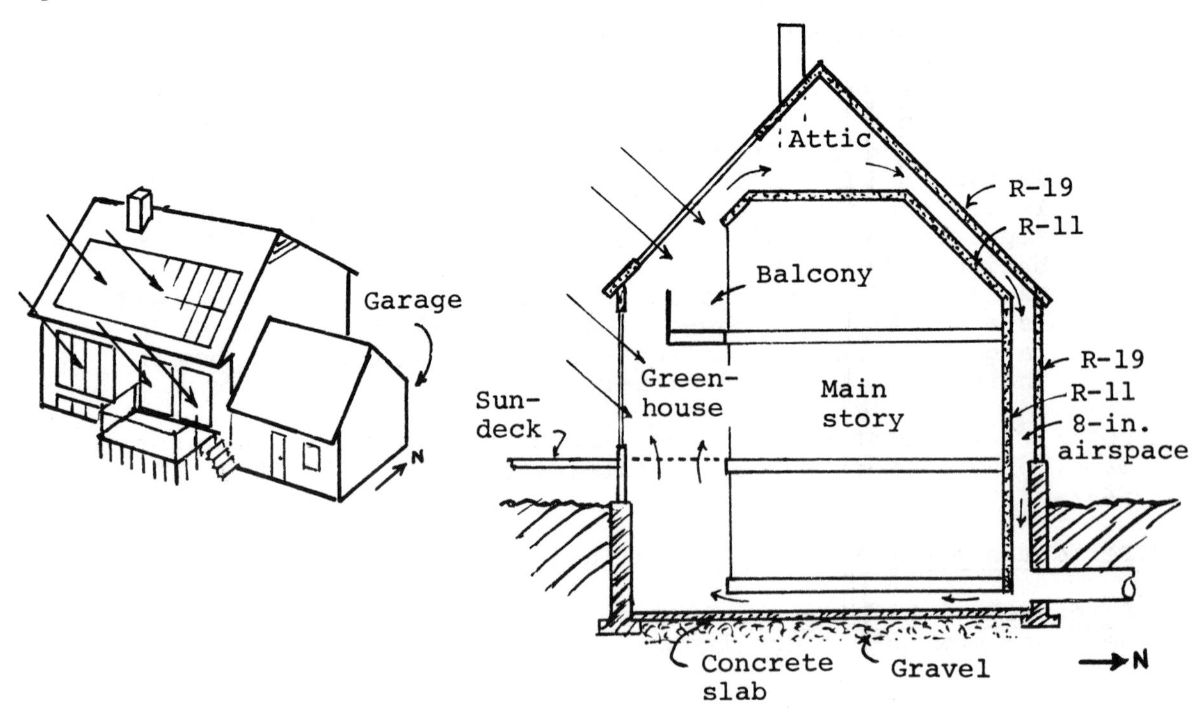

Windows: The total window area is 568 sq. ft., of which 92% (520 sq. ft.) is on the south side, with 12, 24, and 12 sq. ft. on east, north, and west. About 95% of south window area is part of the greenhouse and about 5% consists of small windows serving the basement. All of the windows are double-glazed.

Greenhouse: This is two stories high, 33 ft. x 8 ft. in plan view, and has 520 sq. ft. of glazing. The vertical portion of its south face has a glazed area of 144 sq. ft., consisting partly of slender fixed panels and partly of two large areas leading to a 14-ft. x 10-ft. deck 4 ft. above the lawn. All of this glazing is double and is of tempered glass. The 45-degree-sloping portion includes a 29-ft. x 13 ft. area (377 sq. ft.) that contains two rows of panels with 10 panels per row. Each panel is a standard double-glazed tempered sliding door.

Much of the wall between greenhouse and adjacent rooms is glazed. Some double-glazed sliding doors are used and some single-glazed windows.

The greenhouse floor, on the level of the main story floor, is slotted. The slots are 3/16 in. wide and are 4 in. apart on centers. There are 24 slots in all, and the combined width of the open areas here is 4½ in. In the north central portion of the greenhouse floor there is a 2½ ft. x 6 ft. opening, or hatch, surrounded by a guard-fence. At the east end of the greenhouse there is a staircase leading to the garage door (at about 3 ft. lower elevation) and also leading to the basement. All of these openings and the stairwell facilitate flow of air upward into the greenhouse during sunny days and flow downward into the basement at night or during very cold and cloudy days.

Double-envelope system: The north wall system includes two walls (inner and outer) with an 8-in. airspace between. The above-grade inner and outer walls are 4 in. and 6 in. thick respectively and have nominal resistances of R-11 and R-19 respectively. The airspace is 32 ft. in E-W dimension. The floor joists extend through the airspace but obstruct the airflow only slightly. No dampers to control airflow or block fire-spread are provided. The below-grade outer wall is of 10-in.-thick concrete insulated on the outer face with 2 in. of polyurethane foam. The below-grade inner wall is made with 2x4s and includes 3½ in. of fiberglass providing R-11.

The attic floor is insulated with 3½ in. of fiberglass providing R-11 and the attic ceiling is insulated with 6 in. of fiberglass providing R-19.

The 12-in.-high sub-basement space is almost entirely clear, to permit free north-south flow of air. There are five rows of 8-in.-high concrete blocks here (to support the basement floor beams), but, being parallel to the airflow, they offer little resistance. Although the 8-to-14-in.-high space here is not adequate for use as a crawl space, it is adequate for installing thermometers or flow meters. Access is from a stepdown in the utility room or from base floor area beneath greenhouse. The 4-in. concrete slab that is the floor of the sub-basement space rests on a 12-in.-thick bed of gravel, not insulated.

Hot air from the greenhouse is free to circulate (in a complete loop by gravity convection) into the attic, then northward, then downward in the north wall airspace, into the sub-basement space, thence upward into the greenhouse.

Vapor barrier: There are no polyethylene vapor barriers in the house. However, the fiberglass batts and rolls are paper-backed and have been oriented so that the paper is on the side toward the center of the house. This applies to the house walls and also to the top-story ceiling and the attic.

Auxiliary heat: There is a conventional masonry fireplace. Combustion air is supplied to it via a duct; this air is drawn from the sub-basement-floor space. General infiltration makes up the deficit. In the lowest story there is a moderate amount of electric baseboard heating. There is no furnace.

Domestic hot water: This is of conventional electric type. (A solar heating system may be installed later.)

Cooling in summer: Because the house is on an island, with the ocean nearby (e.g., ½ mi. to the south), little cooling is needed. When cooling is needed, vents in the attic gables are opened manually, with the aid of a rope-and-pulley system; thus hot air is vented to outdoors. Cooler air is drawn into the house (by gravity convection) via a 50-ft. long, 2-ft.-diameter pipe (aluminum culvert) that is buried 7 ft. underground, with an opening 50 ft. north of the house. There is no attic exhaust fan (but one may be added later).

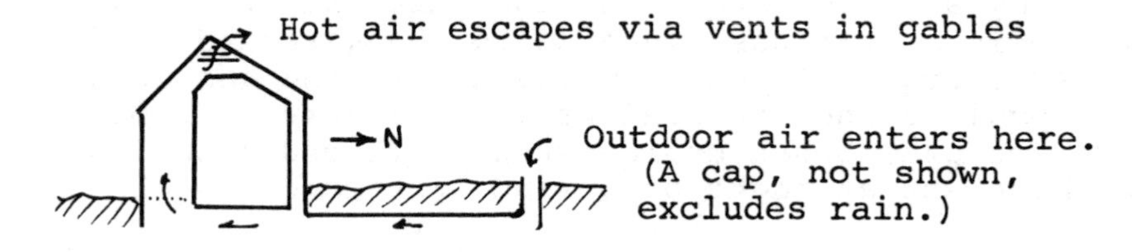

Performance data: By January 1980 little data had become available. The occupants found the house to be very comfortable and have used little auxiliary heat. There has been no excessive build-up of moisture.

References: Popular Science for December 1979, p. 55. Better Homes and Gardens, March 1979. New York Times, Home Section, 9/13/79. Providence Journal, Home Section, 9/16/79. Builder Magazine, January 1980. Also personal communication from occupant and owner.

Note added in May 1980: The performance of this house was extensively monitored in Jan. 1980 by a team from Brookhaven National Laboratory. Preliminary data (but few conclusions) are presented in a report "Monitoring Mastin House for Case Study", by H.T. Ghaffari, R.F. Jones, and G. Dennehy. 21 p. 2/29/80. If my understanding of the data is correct,

the house needs, in January, relatively little auxiliary heat,

even on typical sunny days in January some auxiliary heat is needed

even on January days so sunny that the attic temperature reaches 100°F the crawl space temperature rises only as far as about 52 or 54°F,

at midnight of a typical January day the temperature of a top story bedroom is several degrees higher than the temperature of the living room,

nearly every night in January the temperature in the lower part of the greenhouse falls to near 45°F.

However, these comments give inadequate weight to the many fully successful and attractive features of the house. Also, the report in question was preliminary and may not reflect normal operating conditions. Also, the report contains some obvious errors in data presentation.

Chapter 9

DOUBLE-ENVELOPE HOUSES: OTHER EXAMPLES

INTRODUCTION

Here I discuss other double-envelope houses. In some cases I have fragmentary information only.

CALIFORNIA

Truckee, California
(Near Lake Tahoe, 140 mi. NE of San Francisco)

Dunn House: Deerfield Dr., Lot 12, PO Box 1792, Truckee, CA 95734. 39°N. Altitude: 7000 ft. A 8150 DD location. Tel.: (916) 587-3031. Completed Nov. 1979.

Designer: Built according to general design of the Lake Tahoe House (Smith House), with modifications by builder.

Builder, owner, occupant: Carroll (Rusty) Dunn. (Note: he has built four such houses.) Funding: private.

<u>Building</u>: Two-story, 3-bedroom, 2-bathroom, woodframe house of double-envelope type with crawl space and integral greenhouse. No basement, no garage. Because the house is so similar to the Lake Tahoe House (Smith House), I give only a brief description here. Plan-view dimensions including greenhouse: 36 ft. x 32 ft.; area 2100 sq. ft. Dimensions not including greenhouse: 26 ft. x 32 ft.; area 1700 sq. ft. Living rooms include an office; also a den that opens off the greenhouse. North wall and roof are described below; they are double. The east and west walls are simple (single), insulated to R-19. The south windows (greenhouse windows) are described below. The west, north, and east windows are small, and are covered by curtains at night. All windows are double-glazed. There are no fire-stop dampers in the air-spaces in north wall or roof.

<u>Greenhouse</u>: 36 ft. x 10 ft. Two-story, clerestory, type. Double-glazed with Thermopane. Total glazed area (incl. vertical and 60°-from horiz.): 400 sq. ft. Glazing between greenhouse and living rooms is single and includes some sliding doors. Suspended in upper part of greenhouse there is a 4-ft.-dia. Casablanca-type fan with reversible-pitch blades -- used to drive air upward on sunny days and downward at night. Greenhouse floor consists of 2x6's with ¼-in. slots between to permit airflow. No thermal shades for greenhouse glazing.

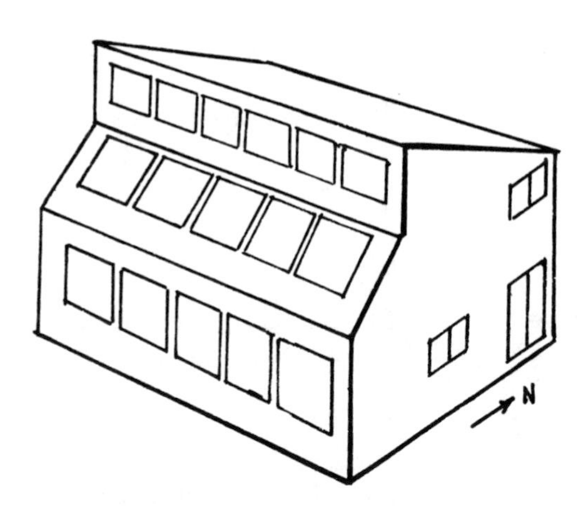

Dunn House: Perspective view

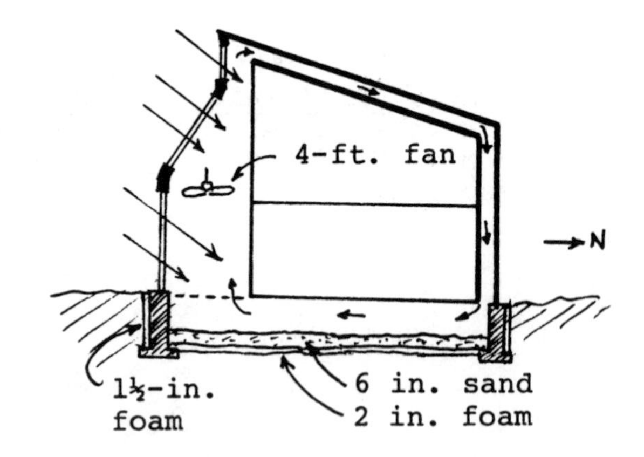

Vertical cross section, looking west.
Not to scale

North wall: There is a 12-in. airspace, or plenum, between the outer component, employing 2x6 studs and R-17 fiberglass, and inner component, employing 2x4's and R-11 fiberglass. Three faces (all except outermost) include sheetrock. No polyethylene vapor barrier is used.

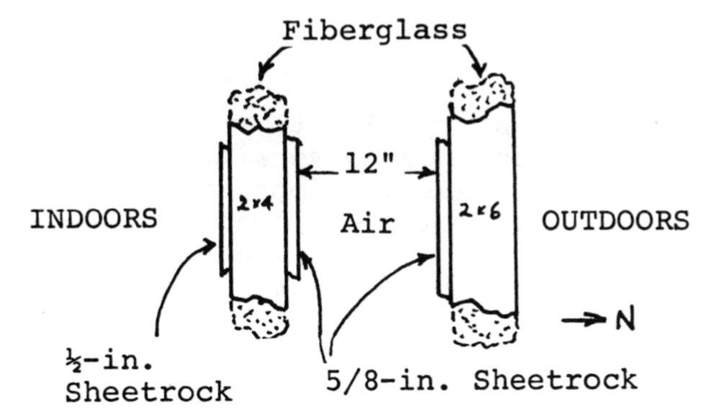

Vertical cross section of north wall, looking west

Roof: This also includes a 12-in. airspace (between 12-in.-deep rafters).

Crawl space: The crawl space beneath first story has 2 ft. of clear height. The 6-in.-thick sand base rests on Visqueen sheeting and this in turn rests on 2 in. plate of urethane foam.

Convective loop: This includes the greenhouse, the 12-in. space in roof, 12-in. space in north wall, and 2-ft. crawl space. Circulation of air upward in greenhouse (during sunny day), or downward, can be assisted by the Casablanca fan situated high up in the greenhouse.

Foundation walls: These extend 2 ft. into the earth. They are completely insulated on outdoor face with 1½ in. of urethane foam. There are five 14-in. x 6-in. conventional vents, closed in winter.

Vents: There are several vents in north wall. Also each bathroom has a vent. The vents are open most of the year. See above for foundation wall vents.

Humidity: No significant humidity problems have been encountered.

Auxiliary heat: This is provided by a wood-burning stove in living room. The flue pipe passes upward through the second story, imparting some heat to it. The house has been wired for some electrical heating, but no actual electrical heaters have been installed.

Domestic hot water: This is heated by conventional means.

Performance studies: Plans have been made for recording temperatures at many locations and for analyzing the resulting performance data.

Source of information: Personal communication from builder-owner.

COLORADO

Denver Area, Colorado

Kern-and-Spivack House: 30 mi. SW of Denver. Exact location and name of owner not indicated. 40°N. Alt.: 8800 ft. 6500 DD. Expected completion: May 1980.

Designer and builder: Kern and Spivack, Inc. of Evergreen, CO.

Cost: (not including land): About $115,000.

Building: 1½-story, 52 ft. x 30 ft., 3000-sq.-ft., 3-bedroom, woodframe, double-envelope-type house with integral greenhouse on south and attached 2-car garage at west. Floor area not including greenhouse: 2500 sq. ft. Main story includes 52-ft-long living-dining room on south and, on north, kitchen, bathroom, laundry, and two bedrooms. Above north region of main story there is a 52 ft. x 15 ft. loft that includes master bedroom, bath, and study. The study is partly cantilevered over the living room; also it has atrium doors opening onto deck above garage. House has air-lock entry.

Windows: South windows are greenhouse windows; see below. Area of west, north, and east windows: 59, 62, and 40 sq. ft. respectively. All are double-glazed.

Greenhouse: 52 ft. x 10 ft. Area: about 500 sq. ft. The area of vertical windows (greenhouse and clerestory) is 366 sq. ft. The area of the sloping windows (at 40° from horizontal) is 108 sq. ft. All glazing is double. Between the greenhouse and the adjacent main-story rooms there is a large area of glass (some fixed, some sliding) and an area of R-11 wall. The greenhouse floor is slotted, to permit airflow; the floor is of 2x6's with ½-in. spaces between.

Double-envelope system: North wall system includes inner wall (2x4's, R-11), outer wall (2x6's, R-19, with 0.006-in. polyethylene vapor barrier), and a 1-ft. airspace between. The crawl space proper is 3½ ft. high; there is R-11 insulation between floor joists above the crawl space, and, below the crawl space, there is a 2-ft. layer of earth resting on a 1-in. layer of Styrofoam. The greenhouse is part of a convective loop. The roof system includes inner roof (2x4's, R-11) and outer roof (2x6's, R-19) with an airspace between, defined by 2x12 rafters.

Auxiliary heating: El Fuego fireplace. Wiring installed for electric baseboards.

Domestic hot water: Five-panels employing oil-type coolant, for solar heating. Back-up heater is conventional.

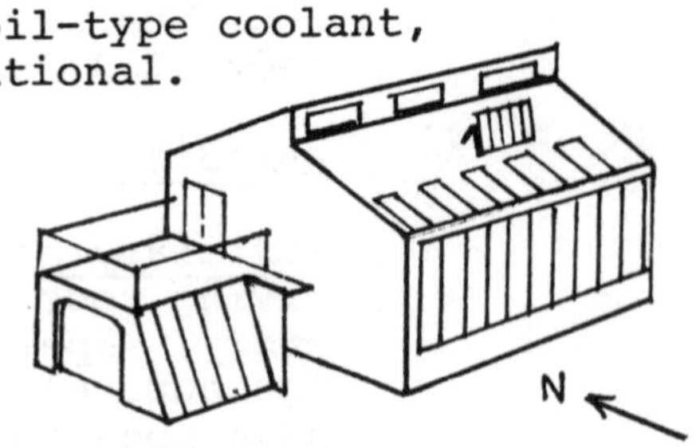

N

Boulder Area, Colorado

Starr House: An interesting variation of double-envelope house has been designed and promoted by Dovetail Press, Ltd., PO Box 1496, Boulder, CO 80306, with the aid of its Housesmiths Journal and its 60-p. booklet "The Solar Starr House Primer". This 1½-story, 1500-sq.-ft., 3 bedroom, 2 bathroom, salt-box type house has air-lock entry at center of south side, with sun chambers to east and west of the entry. Area of south windows is moderate and thermal curtains are provided.

NEBRASKA

Hartington, Nebraska

Demmel House: Rt. 1, Box 234B, Hartington, NE 68739. 42½°N. 1979.

Owner and occupant: Dennis Demmel.

Building: This is approximately of Ekose'a type as exemplified by the Mastin House in Middletown, RI. The houses are similar in many respects.

Performance: With no forced circulation of air, the amount of heat stored in the crawl-space earth was disappointingly small. Trial use of a fan to force rapid circulation greatly increased the amount of heat stored there. In the future such a fan may be used routinely.

In summer, the upper regions of the living space tended to be too hot. The owner believes the situation could be considerably improved by (1) preventing circulation of hot greenhouse air part way down the airspace in north wall (by not having a huge airspace in that wall, but having merely one or two vertical ducts or wells and equipping these with dampers that could be kept closed in summer), (2) eliminating the uppermost glazed area of the greenhouse (or retaining just a small glazed area there to provide solar heating of domestic hot water), and (3) making more use of wind to provide ventilation in summer (inasmuch as it has been found that stack-effect is not very successful in bringing in outdoor air via the big underground cooling tube).

Performance monitoring: By B. Chen et al of the Passive Solar Research Group, University of Nebraska.

Source of information: Personal communication from owner and occupant.

NEW HAMPSHIRE

Bridgewater, New Hampshire (near Plymouth)

Hart House: 8000-DD location. Early 1980.

Architect: Wm. Mead et al of W.M. Design Group.

Owner and occupant: John Hart. Cost: About $65,000.

Building: A two-story, two-bedroom, woodframe, double-envelope house with large integral south greenhouse, basement, crawl space, and (at NW corner) shed. Main structure is 32 ft. x 20 ft. Total floor area: 1550 sq. ft. Living area: 940 sq. ft. Insulation: outer wall, R-19, inner wall, R-11; floor, R-11. Foundation wall insulated with polystyrene foam. Within north wall airspace there are fire dampers controlled by fusible links. Greenhouse, with 550 sq. ft. of glazing (most vertical, some sloping) includes spiral staircase serving small deck at second floor level. At top of greenhouse there is a full-length (32 ft.) grille admitting rising hot air to within-roof airspace.

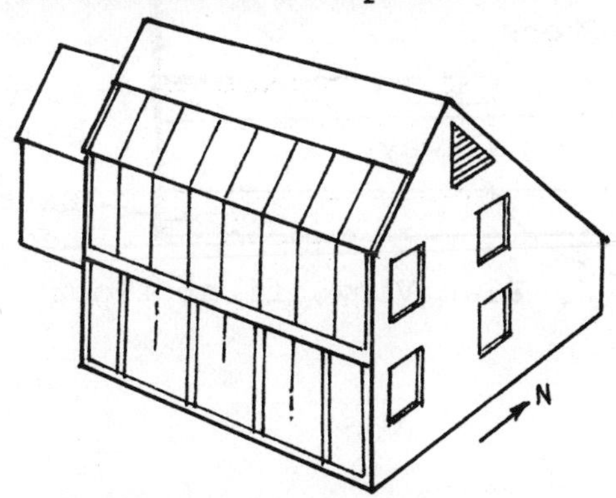

Hart House, Bridgewater, NH, Perspective view.

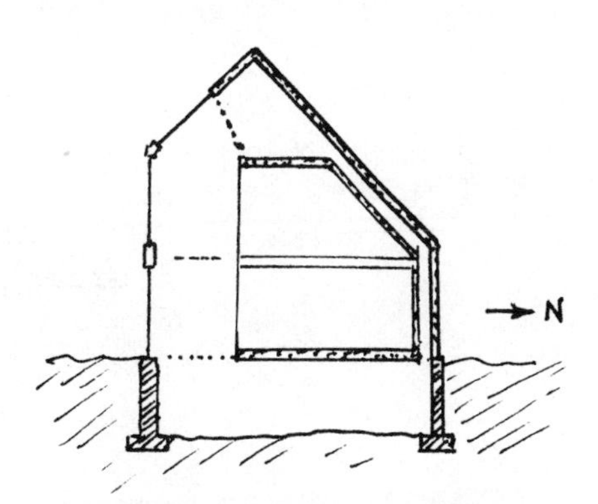

Vertical cross section, looking west

Loudon, New Hampshire (7 mi. NE of Concord, NH)

Beale House: Loudon Ridge Rd., Loudon, NH 43½°N. A 7500-DD location. Occupied Oct. 1979.

Designer: Community Builders, Canterbury, N.H. (Don Booth et al)

Builder, owner, occupant: Galen Beale. Privately funded.

Cost: (not including land): $31,500.

Building: Two-story, 30 ft. x 25½ ft., 1500-sq.-ft., 3-bedroom, wood-frame, double-envelope house with integral south greenhouse, attached 1-car garage and barn on north side, small (not usable) attic space, and crawl space. The main entrance, of air-lock vestibule type, is at NE, near garage. The other outside door is at the west end of the greenhouse.

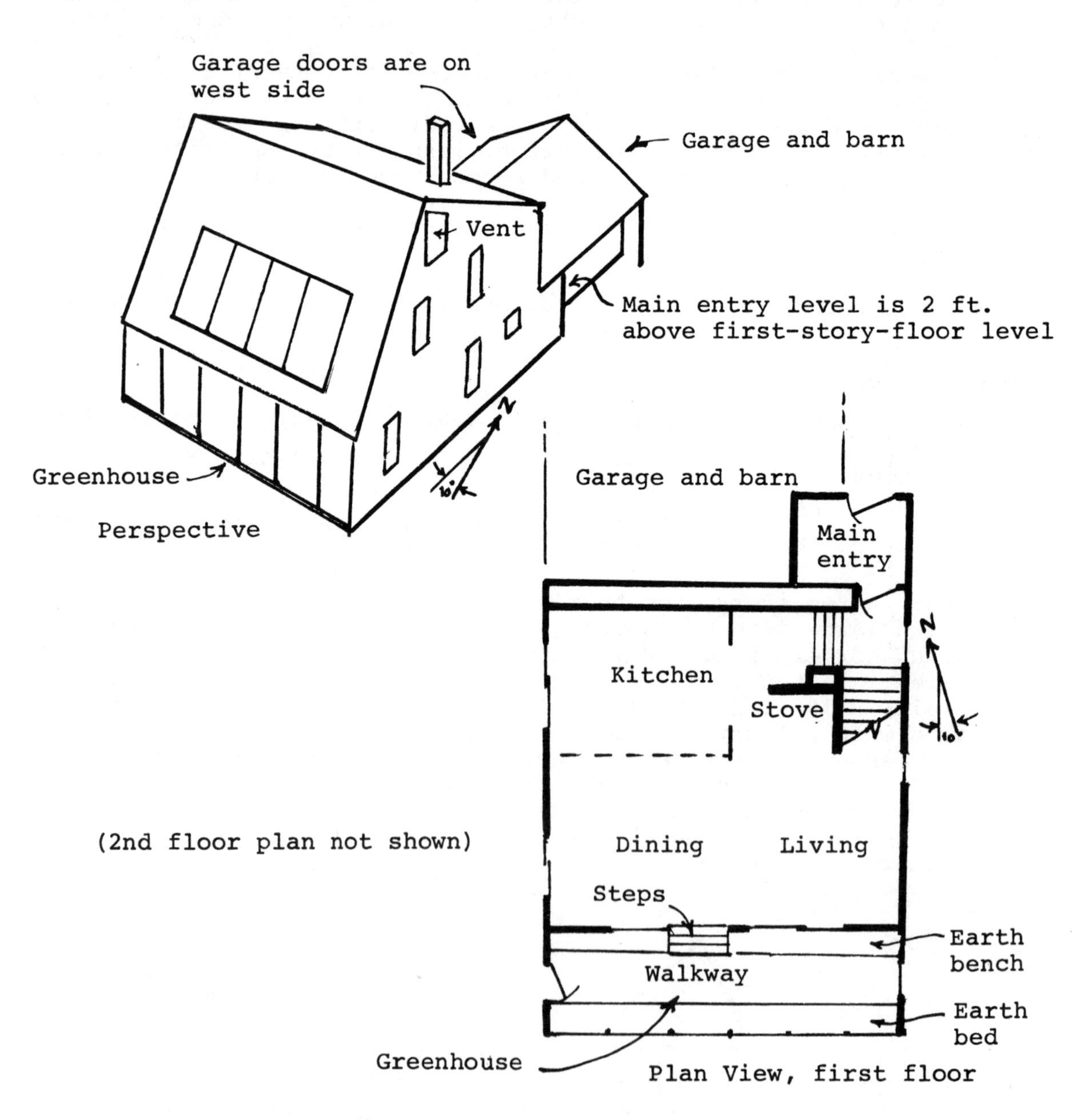

Total enclosed first-and-second-story area of house: 1300 sq. ft.

Total area of first-and-second-story floors within inner envelope space: 1100 sq. ft.

Percent of volume of house sacrificed to (1) upper airspace, not including attic, (2) north wall airspace, and (3) crawl space: about 16%.

Percent of house volume that is within the inner envelope: about 65%.

Windows: Total area of windows of house proper: 427 sq. ft. Of this, about 275 sq. ft. is south window area (vertical and sloping), 36 sq. ft. is on west side, and 36 sq. ft. is on east side. None on north side. All windows (other than a 3-sq.-ft. stained glass window) are double glazed.

Greenhouse: This includes first-story vertical windows and second-story windows at 57° to the horizontal. Total window area: 275 sq. ft. All of these windows are double-glazed. The earth bench and earth bed of the greenhouse are separated by a sunken walkway, which is coplanar with the crawl-space floor; thus the walkway region is a part of the main convective loop.

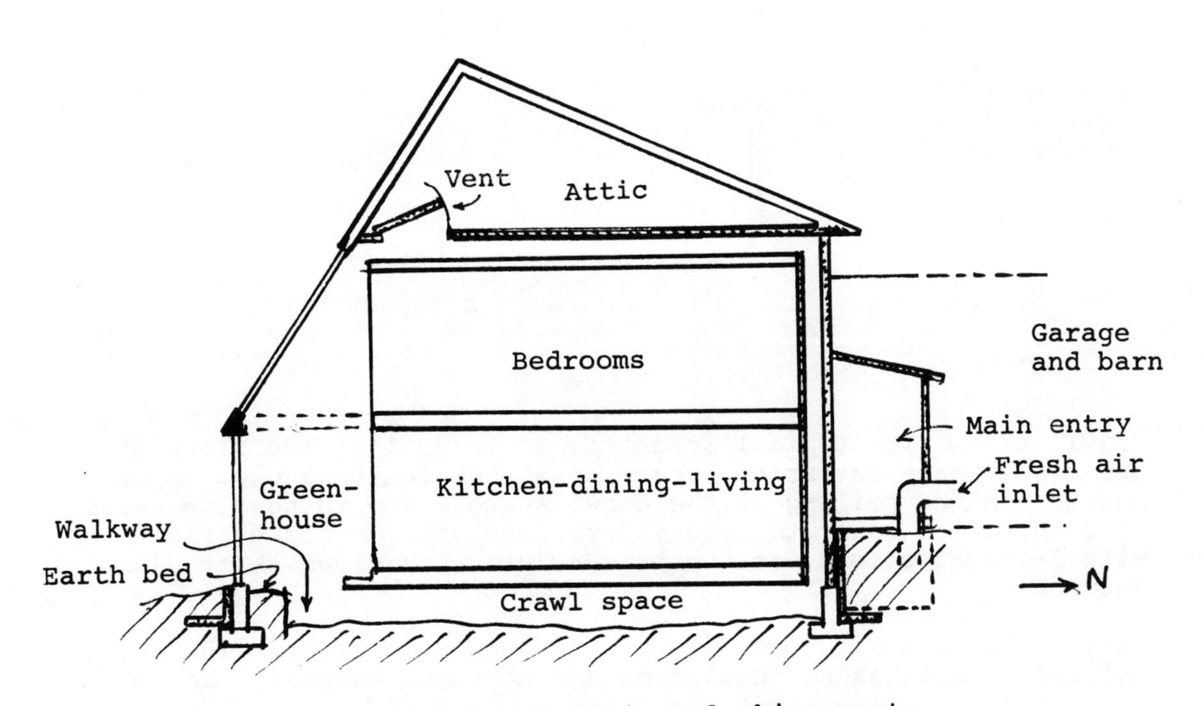

Vertical cross section, looking west

Convective loop: The convective loop includes the greenhouse (discussed above), within-second-story-ceiling space, within-north-wall space, and crawl space. The attic is not a part of the loop.

Within-second-story-ceiling space: This space is 12 in. thick. There is a vapor barrier on the under side of the upper (outer) ceiling. Each of the two ceiling structures employs 2x6's, and each includes 5½ in. of fiberglass. Thus each provides R-22.

Within-north-wall space: This space likewise is 12 in. thick. There are vapor barriers on each side of the space, i.e., on the inner face of the outer wall and the outer face of the inner wall. Each wall includes 2x4 studs, and, on the outer face of the outer wall, there is a 1-in. plate of Styrofoam. Each wall includes 3½ in. of fiberglass; thus the R-values of inner and outer walls are 11 and 16 respectively.

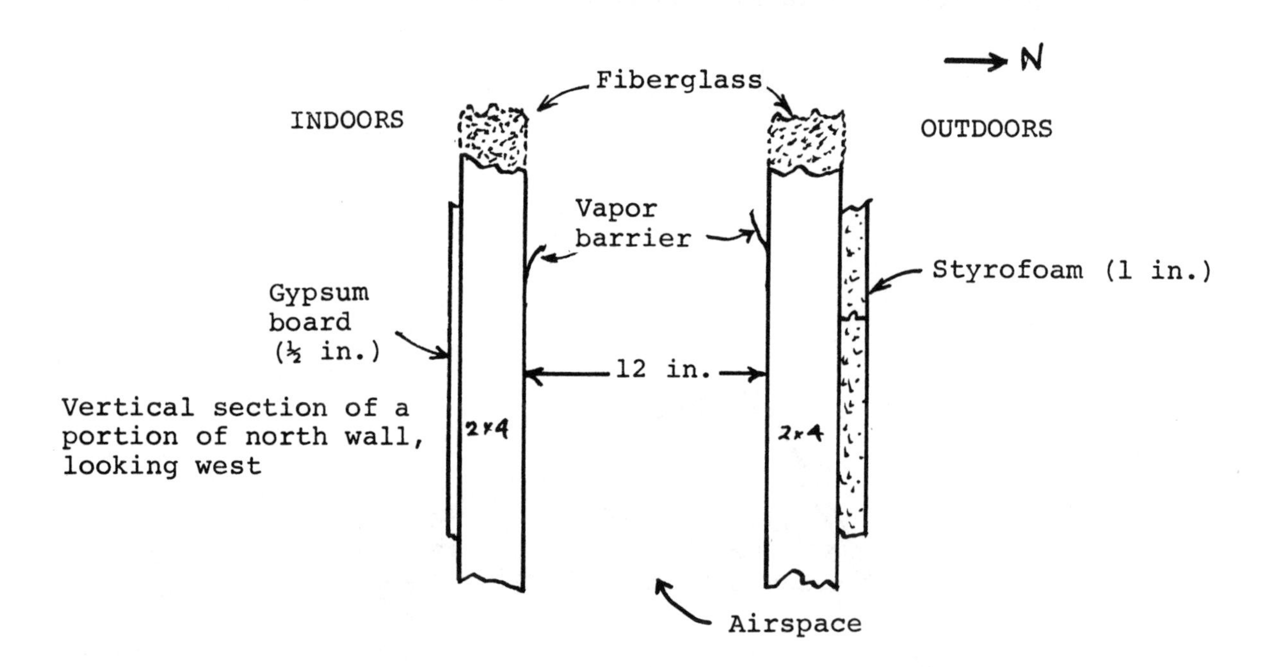

Vertical section of a portion of north wall, looking west

Crawl space: The typical height is 24 to 30 in. The floor of the crawl space is earth. There is no insulation and no vapor barrier in the ceiling of the crawl space. The foundation walls, of 8-in.-thick concrete blocks, are insulated on the outside with 2-in. Styrofoam; at the base a 2-in. Styrofoam skirt flares outward.

Attic: The attic is not part of the convective loop. The attic ceiling (roof) has no insulation and no vapor barrier. In the south part of the attic floor there is an 8 ft. x 4 ft. hatch, controlled by rope and pulley that provides access from the greenhouse. In the east and west gables of the attic there are 12-sq.-ft., manually controlled, vents equipped with louvers and screens.

Vapor barriers: These are of 0.006-in. polyethylene. They are situated on both sides of the north-wall airspace, on the underside of the upper ceiling of the house proper, and on the inner faces of the east and west walls. There is no vapor barrier in the first-story floor or second-story floor.

Auxiliary heat: This is provided by a wood-burning, cast-iron Fisher stove. There is no furnace.

Domestic hot water: This is heated conventionally by electricity. A solar hearing system may be installed later.

Cooling in summer: The vent from greenhouse to attic is opened and the vents from attic to outdoors are opened. Fresh air is drawn into the house via a large-diameter pipe at the north side of the main entry room. There is no conventional air conditioner.

NEW HAMPSHIRE

New Ipswich, New Hampshire (30 mi. SW of Manchester, NH)

New Ipswich House: Timbertop Rd. 43°N. A 7000 DD location. Occupied: 1978. Completed: 1980.

Designer, builder, owner, occupant: Hank Huber.

Building: 36 ft. x 36 ft., 1½-story, double-envelope house (2800 sq. ft.) that includes an integral 36 ft. x 12 ft. greenhouse on the south and, on the west, an attached, 16-ft.-dia. structure of silo (cylindrical) shape. North wall includes inner wall (employing 2x4s) and outer wall (employing 2x6s) with 11 in. airspace between. Ceiling, also with 10 or 11 in. airspace, has R-13 inner component and R-19 outer component. Doors at E and W permit leaving house directly or via greenhouse. Main door is of vestibule type. Very small area of windows on E,N,W. 2½-ft. subfloor space is filled with 2100 hollow-core, 16x12x8 in. concrete blocks oriented so that the cores constitute over 100 channels for gravity convective airflow southward on sunny day (at about ½ to 1 ft/sec) or northward (at much lower rate) on cold night. Such channels, together with greenhouse, within-ceiling space, and within-north-wall space, comprise a complete convective loop. There is no blower. Cost of concrete blocks in 1978: About $1100. There are no thermal shutters or shades. Auxiliary heat: stove burning 2 cords wood per winter, most of this heat being required by the silo, not by the main structure.

Reference: B-415.

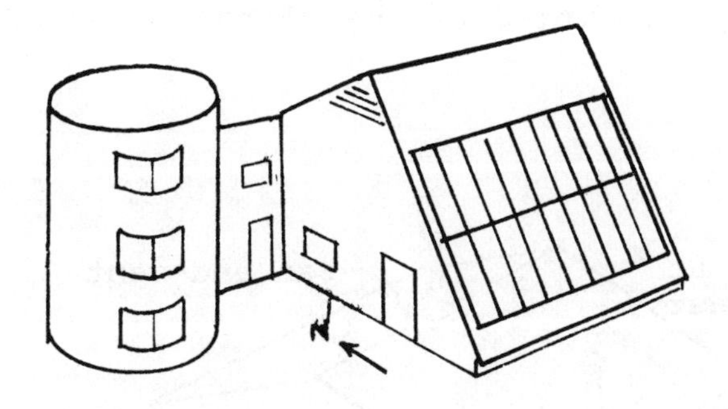

Peterboro, New Hampshire

Peterboro House: Completed in Nov. 1979.

Designer and builder: Hank Huber.

Building: 40 ft. x 34 ft. double-envelope house (2100 sq. ft.).

Peterboro, New Hampshire

Peyton House: Completed in Nov. 1979.

Designer: Hank Huber

Builder: Ned Eldredge

Building: 40 ft. x 36 ft. double-envelope house (2650 sq. ft.).

OHIO

Wauseon, Ohio
(in NW corner of state)

Zumfelde House: County Rd. 12, between D & E, Wauseon, OH 43567. Completed in spring of 1979. 41½°N.

Architect: Dale R. Zumfelde.

Owner, occupant: Paul R. Zumfelde. Privately funded.

Building: This is a 3-bedroom, 3-bathroom double-envelope house with three stories: lower (underground, or basement), main, and upper. At NW corner there is an attached 2-car, 570-sq.-ft., garage. At south there is an integral, three-story greenhouse. Total floor area of house proper (not including greenhouse or garage) is 3160 sq. ft. The house aims 12 deg. E of S.

The lower (underground, or basement) story includes, on south side, master's bedroom, family room, greenhouse earth beds, and, on north side, bathroom and utility rooms. Daylight enters the south rooms via sliding glass doors to greenhouse; no daylight enters from W, N, or E; persons in these rooms can see the greenhouse and some south sky, but have no other view. Between bedroom and family room there is a wood stove. The glass doors and windows facing the greenhouse are single glazed. Near the center of the south wall there is a 12-ft.-long masonry wall to store heat; this wall is shaded in early summer by opaque roof of greenhouse.

The main story includes (at NE) a kitchen and (at NW) a vestibule (leading N to outdoors or W to garage) with bathroom and coat closet. The great majority of the main-story floor area is general family living area; the SE and SW portions of this area have 16-ft.-high ceilings, i.e., are two stories high. Persons in the main story can see into and through the greenhouse, but may fail to see the greenhouse earth beds, which are on the lower (basement) level. In the north central region there is a wood stove. Persons in the SE part of the main story can step out onto a 10 ft. x 10 ft. deck 8 ft. above the level of the greenhouse earth beds.

The upper story includes two bedrooms and a bathroom, in north region, and a play-area in south central region. The SE and SW regions are preempted by the clerestories serving the main story.

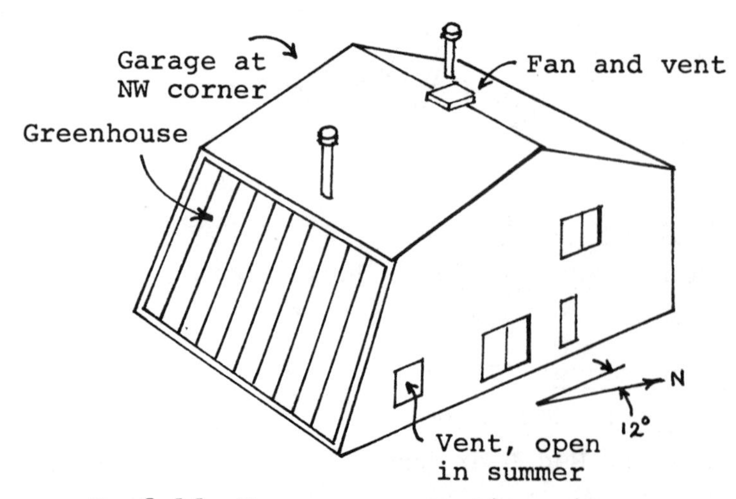

Zumfelde House, perspective view

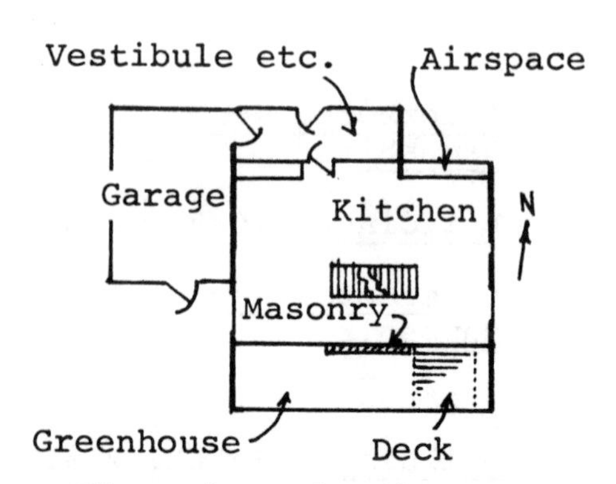

Plan view of main story

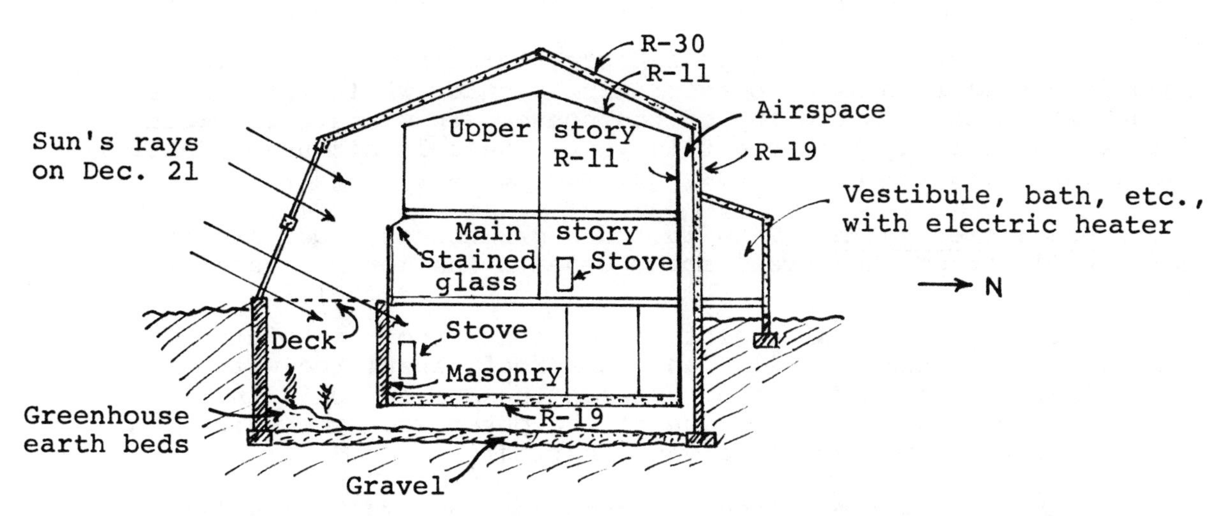

Zumfelde House. Vertical section looking west. Much simplified.

Windows: All windows and glass doors leading to outdoors are double-glazed, and all windows and glass doors from rooms to greenhouse are single-glazed. The window-and-glass-door areas are:

Lower story: south, to greenhouse: 120 sq. ft.; W, E,N,: none.

Main story: south, to greenhouse: 160 sq. ft. vertical and 30 sq. ft. sloping (stained glass) atop masonry wall; W: 50 sq. ft.; E: 65 sq. ft.; N: none.

Upper story: south, to greenhouse: 64 sq. ft.; W: 25 sq. ft.; E: 25 sq. ft.; N: none.

Total glazed area from rooms to greenhouse: 374 sq. ft.
Total glazed area from rooms directly to outdoors: 165 sq. ft.
None of the three bathrooms have any windows. Kitchen has one small window in east wall, but persons in kitchen have unobstructed view to greenhouse. The master's bedroom has no window except to the greenhouse region that is 0 to 8 ft. below grade; it receives much daylight via greenhouse.

Greenhouse: The three-story greenhouse is about 40 ft. long, 8 to 10 ft. wide, and has a plan-view area of 370 sq. ft. It has windows on the south but none on E or W. All of the south windows slope 15 deg. from the vertical and are double-glazed with Thermopane 5/8 in. thick overall (including 1/4 in. airspace). Net area of greenhouse glazing: 540 sq. ft. Gross area: 40 x 16 = 640 sq. ft. There are no thermal shutters or shades.

The greenhouse earth beds are at low-story (basement) level; they cannot be seen from most portions of main story or upper story. About 25% of the greenhouse area is shaded by being beneath a 10-ft. x 10-ft. deck adjacent to the SE portion of main story.

At the east end of the greenhouse there is a vent that is opened in summer to allow outdoor air to enter while hot air is being exhausted from the top of the house. In winter, firewood for stoves is brought in via this vent; the cold air unavoidably admitted enters the greenhouse, not the rooms.

Double-envelope system: This includes greenhouse, roof system, north wall system, and crawl space beneath the lower (basement) floor.

Roof system: This consists of an inner roof (insulated with Styrofoam to R-11), an outer roof (insulated with fiberglass to R-30), and, between them, a 14 to 36 in. airspace. On the underside of the outer roof there is a 0.006-in. polyethylene vapor barrier.

North wall system: This includes an inner wall (insulated with fiberglass to R-11) and an outer wall (insulated with fiberglass to R-19) separated by an 18-in. airspace. The airspace is not bounded on both sides by gypsum board -- but would be if local codes required it. There is a 0.006-in. polyethylene vapor barrier on the inner (S) face of the outer wall.

Crawl space: Extending under the entire area of the lower (basement) story rooms is a crawl space about 24 in. high. The floor of the crawl space consists of a 6-in.-thick bed of gravel (3-in.-dia. stones). The ceiling of the crawl space (floor of the lower story) includes 5½ in. of fiberglass (R-19).

East wall: This is insulated with fiberglass to R-19 and includes, on the warm side, a 0.006-in. polyethylene vapor barrier. Included in this wall are double-glazed windows with a total area of 90 sq. ft. (65, first story, 25, second story).

West wall: This too includes fiberglass insulation (R-19) and double-glazed windows and double-glazed sliding doors with a total area of 75 sq. ft. The two-car garage, adjacent to this wall, is reached via the above-mentioned vestibule.

Foundation wall: Concrete. Insulated on exterior with 2 in. of Styrofoam.

Auxiliary heat: Provided by two wood-burning stoves (in lower and main stories) and three electric heaters in the three bathrooms. Only the heater in the bathroom adjacent to the vestibule, i.e., outside the main living region, is used to any significant extent. Domestic hot water heating: by gas.

Cooling system: Two exhaust fans are used: one at peak of roof to expel hot air via vent, and the other at peak of inner roof to drive hot room air upward into the within-roof space. Outdoor air enters house by vent at E end of greenhouse. There is no underground cooling tube, no air-conditioner. The lower-story rooms have no window or vents (except to greenhouse). Breeze can blow through main and upper stories via E and W windows.

Performance: Said to be excellent. About 1½ cords of wood burned per winter. Small amount of electrical heat used also. Some rooms too hot in August.

References: Personal communications from owner and from architect. See also $10 booklet describing the house in detail; send check to Dale R. Zumfelde, architect, 11 Larkspur Lane, Batavia, OH 45103.

CANADA, BRITISH COLUMBIA

Nelson, British Columbia

Kootenay House #1: 49½°N. A 7200-DD location. Fall of 1979.

Designer and builder: One Step Ahead Energy Systems Ltd., 301 Vernon St., Nelson, BC Canada V1l 4E3. (Allan M. Early, Russell Rodgers, et al)

Performance studies: by Russell Rodgers.

Building: Two-story house of double envelope type. Living area: 1000 sq. ft. Area of integral south greenhouse: 200 sq. ft. There is a partial basement and an air-lock entry (via greenhouse).

Walls: The inner and outer north walls are insulated to R-12 and R-24 respectively. The east wall (likewise the west wall) is simple and is R-24. The inner and outer roofs are R-12 and R-40 respectively. Fiberglass used.

Windows: Windows in east and west walls are double glazed. Those in north wall system are triple glazed: two layers in outer wall and one in inner wall. Windows between rooms and greenhouse are double-glazed with tempered glass. Likewise the greenhouse south windows are double-glazed with tempered glass.

Greenhouse: The floor of the integral, two-story, south greenhouse is of planks spaced to provide 3/8-in. gaps, or slots, to permit upward or downward circulation of air in the main loop. Area of (double) glazing: 255 sq. ft.

Foundations: These are of treated wood and are insulated to R-20 or R-24.

Cooling in summer: Windows at top of greenhouse are opened and some windows and/or doors near ground level are opened also, to permit natural convection.

Auxiliary heat: 5 kW capacity of electric baseboard heaters. Also wood stove.

Performance studies: Detailed performance studies were made by Russell Rodgers. He used simple equipment and analyzed the data with care. His reports (R-205, R-206) are well organized and present sound reasoning. From a quick reading of these reports I draw the conclusions listed below. Warning: I may have misunderstood some of the material; readers would do well to refer to Rodgers' actual reports.)

Horizontal cross section, looking west

In general the house exhibited energy efficiency. Mainly because of the excellence of the insulation and sealing.

Estimated annual gross heat requirement, assuming 68F indoor temperature, is 33,600,000 Btu. Of this, about 25% comes from appliances etc. and 22% from the sun. The auxiliary heating system (e.g., electrical heaters) provided 53%.

Solar energy input was significant but small. In January, at this latitude and at this site near mountains, the sun rises late (10:00 a.m.) and sets early (3:00 p.m.).

On a sunny day little warm air actually reaches the earth.

The effective amount of thermal mass is not great enough to store much heat during a typical day. The massive objects that seem most effective for storing heat are the wood, and also the gypsum boards, of the inner and outer envelopes.

During non-sunny times, much heat is lost via the greenhouse glazing. If the indoor temperature falls below +11F for a fair length of time, the greenhouse temperature drops below 32F. At night the floor of the greenhouse cools to 33F if the outdoor temperature is as low as +7F.

There was little or no evidence of transfer of energy from underside of lowest floor to the earth below by infrared radiation.

The temperature of that earth seems to depend more on the minimum loop temperature than on the maximum loop temperature. The temperature of the earth seems to correlate with the low-temperature portion of the diurnal cycle; at those times greenhouse cold air flows directly across the surface of the earth.

The only significant function of the crawl-space earth seems to be to lessen sudden temperature changes and reduce the chance that the greenhouse will cool down below 32F. After a few days the earth seems no longer capable of performing this function.

Russell concludes that the thermal behavior of this house can be accounted for in terms of standard theories and calculations of heat flow and heat storage.

References: Above-mentioned reports R-205, R-206. Also personal communications.

Chapter 10
DOUBLE-ENVELOPE HOUSES:
DISCUSSION AND CONCLUSIONS

INTRODUCTION

Here I examine closely some of the key features of the double-envelope design. I list the claimed advantages of the system, then I analyze the claims, showing that some are valid, some are of minor importance, and some are of uncertain validity.

I postpone a comparison of the double-envelope design and super-insulation until Chapter 12.

SUMMARY CONCERNING DESIGNS OF COMPONENTS

In general, the double-envelope houses I am familiar with are much alike. Most have two stories; a few have one story or three stories (of which one may be called a basement). Most have about 2000 to 2500 sq. ft. of floor area. Most have greenhouses about 32 ft. x 8 ft., and the glazing is usually double. In west, north, or east windows the glazing is double or triple. Usually there is a crawl space; but in some instances there is a basement instead (or a basement and, beneath it, a crawl space). Main entries are of vestibule type.

There is a within-roof airspace (or within-attic airspace). There is an airspace (as thin as 8 in. or as thick as 12 in.) in the north wall, but there is no airspace in the east or west wall. There are slots or grilles in the greenhouse floor.

Ordinarily there are no ducts or blowers. One house employs a Casablanca type fan in the upper region of the greenhouse.

Typically, only about 50% to 70% of the total volume of the house is kept warm at all times.

Typically, if a person in a living room wishes to walk to the south lawn, he must walk through the greenhouse. In doing so, he may have to operate two sets of sliding doors: a set between greenhouse and room and a set between greenhouse and lawn.

I believe most people would call most of these houses highly attractive, interesting to study, and all-around pleasant places to live.

ADVANTAGES CLAIMED FOR THE DOUBLE-ENVELOPE DESIGN

From discussions with designers, builders, owners, and occupants of double-envelope houses, I gather that they see these advantages:

1. Most of the heat needed comes from the sun. Little auxiliary heat is needed.
2. The rooms are extremely comfortable -- because (a) their walls are bathed in fairly warm air circulating in the within-wall spaces and accordingly wall temperature is practically as high as room air temperature, (b) there are no jets of cold air entering the rooms (even if cold jets enter the within-wall spaces), and (c) humidity is near-ideal (thanks to spread of water vapor from the greenhouse).
3. The relationship between south rooms and greenhouse is highly pleasing. When the sliding doors or the windows between greenhouse and south rooms are open, fragrance of the greenhouse plants can flow directly into the rooms, providing a "Hawaii-like atmosphere".
4. The greenhouse has enough area so that considerable quantities of vegetables or flowers can be grown there throughout a large fraction of the winter.

5. Excessive heat in the greenhouse is transferred, purely by gravity convective airflow, to a large thermal mass in the crawl space. Thus the greenhouse does not become intolerably hot on sunny days and the crawl space stores much heat on those days.

6. On cold nights, heat from the thermal mass in the crawl space flows to (a) the greenhouse (preventing it from becoming too cold -- colder than about 40°F or 45°F) and (b) crawl-space air that circulates into the within wall spaces provides some thermal protection to the rooms.

7. Construction of the double-envelope is reasonably simple, reasonably inexpensive.

In summary, the design provides, at low cost, many superb features -- according to the proponents.

SOME COUNTER-ARGUMENTS

Many designers, architects, and builders of solar heated houses of common types appear to agree with many of the above-listed arguments, but remain unconvinced by some of them. They express themselves along these lines:

1. Everyone knows that integral greenhouses can be a great pleasure and can supply much heat to the rooms. But many well-known solar houses have integral greenhouses and supply much heat -- even although they by no means conform to the double-envelope design.

2. In any well-insulated, reasonably airtight house the occupants may be fully comfortable. Walls are warm. There are no cold jets of air. If the humidity is too low, a small humidifier may be used.

3. Linking the greenhouse and rooms (or greenhouse and convective loop) may create a pleasant atmosphere. Yet it may pose the risk that bugs, insects, sprays, etc., may spread from greenhouse to rooms. If the greenhouse is especially humid, the rooms may become too humid. When the greenhouse is very cold, the south glass doors and south glass windows of the contiguous room may become somewhat cold, impairing comfort.

4. Although large quantities of vegetables can be grown in an integral greenhouse, in some instances the greenhouse is largely ornamental. Also much greenhouse floor area may be preempted by the longitudinal walkway, the crosswalks to the outdoors, the entry-ways or staircases, and the slots, grilles and hatches. In mid-winter the greenhouse is likely to be too cold to grow vegetables.

5. Despite flow of hot air from the greenhouse to the within-wall spaces, the greenhouse may become uncomfortably hot. Also, we cannot expect very much heat to be delivered to the crawl space, and we must expect that the amount of heat actually finding its way into the earth beneath the crawl space to be small. Thus the overall efficiency of transporting and storing heat may be low. Anyway, there are better ways of transporting heat and better ways of storing it.

6. Luke-warm air from the thermal mass may indeed keep the greenhouse from becoming very cold. But such air serves no useful function as regards the room (or only a very minor function).

7. The added construction cost must be expected to be considerable. Building the two roofs, the two north walls, and the sub-floor space, and taking proper precautions against fire and humidity, and providing adequate access for inspection and servicing, involve considerable extra material and extra labor. Installing windows in a double wall is some trouble.

Such critics may point out also that:

Much of the gross volume of the house (perhaps 15% or 30%) may be preempted by the airspaces. The volume that is at all times kept warm may be only 50% to 70% of the total. In this respect the house may be smaller than one might have assumed, and the cost per continually-heated-square-foot (or cubic foot) may be larger than expected.

The permanent interposition of greenhouse between south rooms and south lawn may sometimes be regretted. To reach the lawn, an occupant may have to walk across the greenhouse, opening and shutting doors en route. The greenhouse becomes, to some extent, a traffic area.

The permanent "marriage" of greenhouse and best (south) rooms may be a mixed blessing. If the greenhouse is much too hot or much too cold, this condition may be, to some extent, compulsorily shared with the rooms. Likewise if the greenhouse is sometimes extremely moist, or if it sometimes contains many insects, the rooms may bear some of the brunt. (Should there be screens in addition to doors between greenhouse and living rooms? If so, this adds to the difficulty of walking back and forth from rooms to greenhouse.)

Heat-loss through the huge windows of the greenhouse may, on cold nights, be so great that the occupants may decide to install thermal shutters or shades (costing, say, $4/sq. ft., or about $1500 in all). Operating them each night and each morning is a bit of trouble.

If there are no windows in the north wall (or none in an east or west wall), interesting outlooks, or views (e.g., summer sunsets) may be lost. Also, cross-ventilation on summer days or cool summer nights may be sacrificed.

Squirrels, raccoons, cats, dogs, birds, etc., may find their way into the within-wall airspaces, or sub-floor space, or attic space -- since these spaces are directly joined to the greenhouse. Such animals may die. May smell. Children may throw toys into these spaces, and retrieving the toys may be difficult. Large portions of the airspaces may be practically inaccessible: hard to see into, hard to climb into.

FOCUS OF OBJECTIONS

The focus of objections is, I think:

the within-wall airspaces and, in general, the provision of an all-the-way-around convective loop,

the reliance on gravity convection for delivering heat to the sub-floor storage.

There is no quarrel with having a greenhouse, or having excellent insulation in roof and walls, or having only small window areas on west, north, and east, or using double or triple glazing, or employing vestibule-type entrances, or of making use of the excess heat in the greenhouse.

OBJECTIONS TO THE WITHIN-NORTH-WALL AIRSPACE

Consider a typical within-north-wall airspace: it may be 10 in. thick, 16 ft. high, and 32 ft. in E-W dimension.

As a plenum for airflow the airspace may be slightly handicapped by the joists, ties, etc., that extend through it -- and by any major interruptions for windows or entrance door.

Installation of sheetrock or equivalent on both sides of the airspace may be necessary for fire protection. Also, provision of fireproof dampers and fusible links may be necessary.

Installation of vapor barriers (or equivalent) on both sides of the airspace may or may not prove necessary. It is essential that little or no condensation of moisture on the insulation or on the studs etc. occur.

Animals may find their way into the airspace. May die there. May smell.

Making inspections and repairs in the airspace may be difficult.

The within-north-wall airspace takes up about 3% of the gross volume of the house, or about 4% or 5% with respect to the continually heated volume. Perhaps this corresponds to 2% or 3% of the total cost of the house -- perhaps it corresponds to $1000 or $2000. (Note, however, that in a great many types of houses there is considerable waste space.)

OBJECTIONS TO THE WITHIN-ROOF AIRSPACE

The arguments are much like those listed above. One may worry about added construction cost, protection against fire and moisture, penetration by animals, access for inspection and repair, and waste of a few percent of the total volume of the house.

OBJECTIONS TO THE SUB-FLOOR AIRSPACE

The arguments are similar. There are also these worries:

Excessive moisture from ground water, especially in the spring or after a many-day rainstorm.

Snakes, rats, moulds.

Household pets could crawl into such space and adopt it as a toilet.

Crawling infants could crawl into such space.

While I may be overstating some of the drawbacks (and, to date, most of my worries are without foundation), I tend to take a dim view of providing spaces that are accessible to birds, rodents, etc., and to children, yet are inaccessible (or nearly so) to adults. Perhaps at the very least the spaces should be closed off at each end by ½-in.-mesh galvanized steel screens. Unfortunately such screens would considerably obstruct airflow, I suppose.

COMMENT ON THE PURPOSE OF THE WITHIN-WALL AIRSPACE

Is the purpose to keep the wall-surface-toward-room warm? If so, why do not the designers provide spaces within the east and west walls also? (See later section concerning temperature of walls of room.)

Is the purpose to intercept in-leaks of ice-cold air? If so, why not provide such spaces within the east and west walls also? (See later section concerning the "interception-of-cold-air" fallacy.)

Is the purpose to provide a path for flow of hot greenhouse air to the crawl space? If so, why not use a direct, much shorter, route? Why not use a blower and duct, the duct running directly from the upper part of the greenhouse into the crawl space?

Is the purpose to provide a path for nighttime flow of warm crawl-space air to the greenhouse? If so, why not provide a much shorter broader path, namely a path via the grilles of the greenhouse floor? If there are several grilles, air can flow upward through some and downward through others. Why construct a circuitous, 75-ft-long path instead of a direct 15-ft-long path?

THE MOST CONTROVERSIAL CLAIMS

The most controversial claims of the double-envelope proponents are these:

Claim 1 On cold nights the rooms are especially comfortable because their walls are bathed in the circulating luke-warm air.

Claim 2 On cold nights the greenhouse is kept well above freezing -- kept above about 45°F.

Note: these alleged benefits are contingent, obviously, on (a) the sunny-daytime collection of much solar energy by the greenhouse and (b) the transfer of a fair fraction of the collected energy to the thermal mass beneath the crawl space. These prerequisites are discussed in a later section.

Analysis of Claim 1

In analyzing the claimed benefit of bathing the walls of the room with luke-warm air, it is helpful to consider these key facts:

A. The bathing process applies to some rooms, but not all. In the rooms to which it applies, it applies to only one or two of the bounding surfaces, not to all six. The walls of some of the most important rooms (lower south rooms) are not bathed at all -- except by greenhouse air, which may sometimes be as cold as 45°F.

B. Typically, the effect of bathing a given wall is to raise the temperature of the surface toward the room by only about ½ F deg.

To show that this is so, consider, first, a simple, 8-in.-thick, R-30 wall with indoor and outdoor air temperatures 70°F and 30°F. The temperature of the wall surface that is toward the room is less than 70°F -- because of the resistance (about R-0.67) of the stagnant air film adjacent to this surface. Because 0.67 is only about 2.2% of the overall R-value of 30.67, the temperature drop between room air and the wall surface in question is only about (2.5%)(70 - 30) = 0.88 F deg.

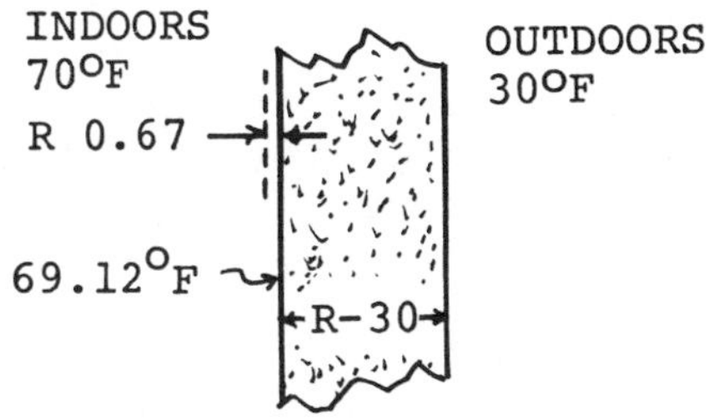

INDOORS 70°F
R 0.67
69.12°F
R-30
OUTDOORS 30°F

Now consider a pair of walls, each 4 in. thick (R-15), with an intervening airspace in which 60°F air is circulated. Here, the overall delta T across the inner wall is only about 10 F degrees, and the 0.67 R-value of the stagnant airfilm is 4.3% of the overall R-value of 15.67. Thus the temperature drop across the airfilm is (4.3%)(10) = 0.43 F deg.

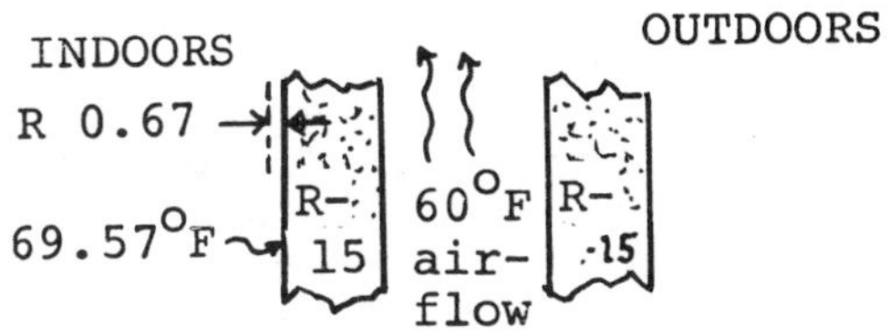

INDOORS
R 0.67
69.57°F
R-15
60°F air-flow
R-15
OUTDOORS

Conclusion: Changing from (a) a single 8-in. wall to (b) a double-wall with 60°F air in the intervening airspace raises the temperature of the wall surface that is toward the room by only 0.88 - 0.43 = 0.45 deg. This, then, is the benefit: an increase of about ½ degree in the wall temperature.

By way of contrast, consider a room having walls consisting of a single layer of 3/4-in. pine board. Here the R-value of the stagnant airfilm (0.67) is roughly comparable to that of the board (about 0.8). Thus if indoor and outdoor temperatures are 70°F and 30°F, the temperature of the board surface toward the room is roughly midway between these temperatures, i.e., is about 50°F or 55°F. This is bad! It makes for discomfort. There is much need for improvement. (But, equally clearly, the single pine board has no resemblance to a house-wall that contains 8 inches of fiberglass.)

C. If only one of the six boundary surfaces of the room is served by a bathing operation, the average effect per boundary surface is one sixth as great, i.e., less than 0.1 F deg. If only two walls are served, the average effect is less than 0.2 deg.

D. If the occupants are fully clothed, so that there is a radiative buffer between their skin and the room walls, a 0.1 or 0.2 F deg. change in average wall temperature is trivial. Even a tenfold greater change would be trivial or near-trivial. (I mean: trivial compared to the many-full-degree temperature changes that occur from hour to hour in most houses, and the ceiling-to-floor temperature differences, and trivial compared to the occupants' preferences which vary from time to time depending on what the occupants are doing.)

As regards the south rooms, there is this additional circumstance: a large fraction of the south wall of such room is single glazing, and, on cold nights, the air immediately south of this glazing may be at about 50°F or 60°F. Because the R-value of the stagnant film on the north side of this glass is comparable to the R-value of the glass itself and the airfilm on the other side, the temperature of the glass surface toward the room may be several degrees below room temperature. In other words, the deficiency of surface temperature here, on cold nights, may far outweigh the slight increment in surface temperature of some other bounding surface. (If shades are drawn, the argument is not applicable.)

All of the foregoing arguments may be replaced by the single well-known observation that, in a house having walls that include a single fiberglass layer at least 6 in. thick, the wall surfaces toward the room are extremely close to room air temperature and do not constitute any appreciable discomfort.

Another bit of accepted knowledge is that, in such a house, unshuttered windows may be the main source of thermal discomfort (if any) and infiltration of cold outdoor air may be a significant source of discomfort. Efforts to shutter the windows at night and to stop in-leak of cold air may well be worthwhile. But there is no justification for worrying about wall temperature. In any well-insulated, reasonably airtight building wall temperature is not a problem.

Analysis of Claim 2

In analyzing the claimed benefit that, even on cold nights, the greenhouse temperature is kept from dropping below about 45°F, one finds that the questions reduce to:

A. What, actually, is the airflow pattern between greenhouse and crawl-space thermal mass?

B. How much heat is actually stored, on a sunny day, in the crawl-space thermal mass?

C. Which thermal masses supply on cold nights, most of the heat needed?

Answer to A: If the thermal mass beneath the crawl space is warm, much heat will surely be supplied by it on cold nights: much heat will be supplied to the greenhouse. Warm air will flow from crawl space to greenhouse, and cold air will flow from greenhouse to crawl space.

Depending on the detailed design of the double-envelope house in general and the convective loop in particular, different patterns of nighttime airflow may be expected.

Counterclockwise flow: If all of the airflow passages (including the greenhouse itself) are slender and roughly equally slender, the airflow is likely to be counterclockwise, judged by a person looking west -- see sketch. Why? Because the outer face of the greenhouse is far more poorly insulated than the outer wall of the north wall system, and accordingly the greenhouse loses heat especially rapidly and its air becomes especially dense. Therefore gravity produces a counterclockwise flow -- air flows downward in the greenhouse, upward in the north wall airspace.

Nighttime airflow in convective loop. Vertical cross-section, looking west

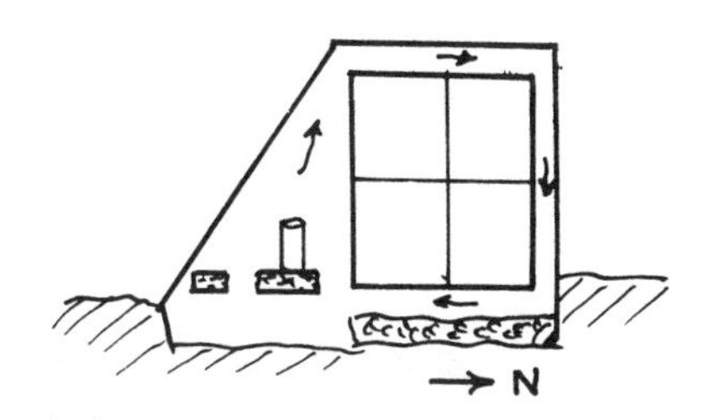

Clockwise flow: Suppose that the greenhouse includes a very large thermal mass -- includes, say, much earth or many water-filled drums. Then it may be that, despite the low R-value of the big window area, the greenhouse will cool down more slowly than the air in the within-north-wall airspace. In such case, the greenhouse air will be relatively low in density and therefore gravity will tend to produce a clockwise flow.

Two-way flow through greenhouse floor: If the greenhouse floor has a very large area, and contains several large grilles, there may be two-way flow through the greenhouse floor. See sketch. Within the convective loop as a whole there may only be trivial flow.

Which flow pattern will occur ordinarily? I do not know. I suppose that the answer depends on the sizes (pneumatic conductances) of the various passages, the relative amounts of thermal mass in greenhouse and in the north wall system, the relative surfaces areas (this controls how easily the heat can penetrate into, or pass from, the thermal mass), and the transient conditions of start-up.

I believe the flow pattern is bi-stable: within certain limits, if the flow starts off counterclockwise, it will continue so; or if it starts off clockwise, it will continue so. Relative temperatures of crawl-space thermal mass, greenhouse thermal mass, and north-wall-system thermal mass may be important.

Whether the height of the crawl space is adequate to permit simultaneous two-way flow may be important. I mean, for example: southward flow 1 in. above the crawl-space floor, and northward flow 10 in. above it.

Another controlling factor may be the rate at which heat is supplied by the rooms. The south rooms may supply much heat to the greenhouse at night inasmuch as, between these rooms and the greenhouse, there are large areas of single-glazed windows and sliding doors. The amount of heat supplied via the inner north walls to the within-north-wall airspace is much less, in view of the thick intervening insulation. Such considerations tend to favor a clockwise airflow pattern.

Changes in speed and direction of wind, and changes in amount of solar radiation at the end of the day, may play a role in determining which airflow pattern prevails.

Which airflow pattern do the designers wish to prevail at night? My impression is that they wish the pattern to be counterclockwise, i.e., the reverse of the sunny-day pattern.

Is there good reason for such wish? I do not know.

Perhaps this is a good question: What fraction of the time (on cold nights) will the pattern be counterclockwise? What fraction clockwise? What fraction two-way-through-greenhouse-floor?

Other questions are: Will the pattern be the desired one when heat is most badly needed in the greenhouse? What steps can the occupants take to make sure that the desired pattern prevails? What are the penalties if a wrong pattern prevails? Or can it be that "any old pattern will do"? What about speed of airflow? Could it happen that the pattern is right but the speed is wrong?

Perhaps I am overstating the uncertainties or difficulties. Perhaps the various performance tests underway will show the patterns and speeds of airflow to be satisfactory under a wide range of conditions. But, at present, I feel uneasy about a phenomenon that is counted on to provide an important function, yet can be affected by so many circumstances -- some of which the designers do not understand and some of which are beyond the ability of the house occupants to control.

Answer to B: How much heat is actually stored, on a sunny day, in the crawl-space thermal mass? I do not know the answer. A later section suggests that the answer is: only a small -- perhaps very small -- fraction of the surplus heat in the greenhouse. Making heat flow by gravity-convective flow to a location lower down than where the air is heated is necessarily inefficient. Requiring the air to flow around two sharp corners may greatly slow the flow.* Bringing the warm air in the crawl space into intimate contact with the earth beneath the crawl space (rather than allowing it to "hug the ceiling" of the crawl space) may be difficult and/or impractical. Transfer of energy, by radiation process, from warmer crawl-space ceiling to cooler crawl-space floor (earth) is a slow process. The earth itself, if dry, is a poor thermal conductor; most of the heat delivered to the earth may reside in the uppermost few inches of the earth. If the earth is moist, evaporation may occur, producing some cooling (not desired).

---

* I have heard the suggestion that all corners along the airflow path should be gently rounded. Alternatively, turning vanes could be used.

Answer to C: Which thermal mass supplies (on a cold night) most of the heat? The crawl-space thermal mass? Or the thermal mass in the north wall system? Or the thermal mass of the greenhouse earth, floor, and walls?

I suspect that there is no general answer.

The mass of a 6-in. layer of earth beneath the crawl space is far greater than the masses of the north wall system or of the greenhouse and its components. But most of that earth is insulated, e.g., by the topmost inch of earth. What counts is the mass, the specific heat (i.e., per pound), the exposed surface area, the thermal conductivity, the rate at which air flows past, and the extent to which the airflow is turbulent. It might happen that, during the first hour of airflow, a certain component (say the sheetrock within the north wall system) contributes the most to heat uptake (or heat output), but in a subsequent hour some other component makes the maximum contribution.

It may be that, during the first hour, the earth beneath the crawl space is of main importance. Perhaps, during some later hour, the surface of the earth will have cooled sufficiently so that some other element, such as the ceiling of the crawl space, will be of main importance. Perhaps at some stage the thermal mass of the greenhouse itself will become of main importance.

Perhaps various monitoring groups will have firm answers soon.

Summary regarding analysis of Claim 2: It is clear that, on cold nights, heat will indeed be supplied to the greenhouse air. But:

> There are many possible airflow patterns; I do not know which one, or which ones, predominate, or which is most effective, or how the patterns may be controlled in practical manner.
>
> I do not know how much heat will be stored, on a sunny day, in the crawl-space thermal mass; but I suspect that the amount is either small or very small.
>
> Of the heat supplied to the greenhouse air, I do not know which of several thermal masses makes the main contribution, or how the contributions will change under various circumstances.

CLAIM THAT HEAT IS SUPPLIED BY DEEP-DOWN EARTH

Some proponents of double-envelope houses or other houses that "sit deep" in the earth claim that, on cold nights in winter, heat is supplied to the house by deep-down earth. If the very-deep-down earth tends to stay at a uniform temperature of, say, 45F the year around, then if on cold winter nights the crawl-space air or greenhouse air tends to become colder than this, the deep-down earth could indeed supply some heat. For example, it could supply enough heat so that the greenhouse would not cool down below 40F.

Obviously, the uppermost earth of the crawl space can supply heat on such occasions. It has a buffering, or thermally smoothing, effect. No one doubts this. But most of the heat in question was supplied to this earth from above.

Many people have the notion that, at least in the coldest part of winter, the very-deep-down earth makes a contribution. They believe that heat from earth that is 10 or 20 ft. down flows (very slowly) upward, and has a helpful effect.

Grim fact: Ordinarily, there is no such effect. The very-deep-down earth beneath the center of the house sends no heat upward; rather, there is a continuous, year-round, downward flow here.

To prove this fact, one may employ a thought experiment. Suppose that there are several thermometers buried 12 ft. deep beneath the crawl space, and some other thermometers buried 11 ft. deep. Then if the 11-ft-deep thermometers always read higher than the ones at 12 ft., we have positive proof that the heat-flow in that neighborhood, i.e., at that depth, is always downward. But will the 11-ft.-down temperature always be higher than the 12-ft.-down temperature? A little thought shows the answer to be yes. The temperature 11 ft. down changes extremely slowly, extremely little; it remains very close to the year-around average temperature there. All summer, with the house rooms at 70F to 90F, the crawl-space air tends to warm up; during the winter that air is typically between 40F and 60F. Clearly the annual average temperature of the air here is between 50F and 70F. Accordingly the temperature 11 ft. down will tend to be between the natural, very-deep-down, temperature (45F) and the 50-to-70F temperature. This completes the proof. Every day, and indeed every hour throughout the year, heat is flowing downward from the 11-ft. depth to the 12-ft. depth. At no time is there an upward flow of heat here. Very-deep-down earth contributes nothing to keeping the house and greenhouse warm.

Even in a southern state where the deep-down, year-around temperature is 60F, it is hard to imagine circumstances that would make the temperature 11 ft. down lower than the temperature 12 ft. down. Here too the conclusion is that the direction of heat-flow at great depths is always downward (beneath the house).

Corollary 1: It would help, in winter, if there were a horizontal layer of insulation several feet down in the earth beneath the crawl space. The insulation would slow the steady downward flow of heat. The house and greenhouse would stay warmer.

Corollary 2: All of the heat supplied to the house by the crawl-space earth is heat that has previously been delivered to this earth from above. Thus if little heat is delivered to this earth from above during warm spells, little will be given out during cold spells. Mother Earth is no thermal Santa Claus.

THE CONTENTION THAT "IF IT WORKS, YOU DON'T NEED TO KNOW HOW OR WHY"

It is certainly true that if much heat is stored during sunny hours and much heat is supplied to the greenhouse on cold nights, the house occupant is happy and does not need to know just how or why the system functions. Up to a point, this contention is justifiable.

The trouble is that if the designers lack understanding, their future designs -- if different from today's -- might function poorly. Slightly different designs of beds of earth or rock, or of within-wall spaces, or of the lining sheets of such spaces, or of the insulation of such spaces, might produce unexpected drops in system efficiency. Likewise changes in greenhouse size or shape, or changes in greenhouse thermal mass, or changes in house orientation, or changes in climate, might hurt system performance. Designing and building without an understanding of the physical mechanisms involved is a risky business.

CLOSE LOOK AT "BATHING THE NORTH WALL"

If the air circulated (on a cold night) in the within-north-wall space is at 60°F and the air in the north rooms is at 70°F, then:

When outdoor temperature is 0°F, the circulating air
- helps keep those rooms warm;
- loses much heat to the outdoors (heat-loss is much greater than if the circulation were stopped and the air here were allowed to cool down to its equilibrium temperature of about 35°F).

When outdoor temperature is 50°F, the circulating air
- accomplishes nothing (room comfort -- and heat-loss to outdoors -- same as if the circulation were stopped).

When outdoor temperature is 60°F, the circulating air
- slightly hurts room comfort (makes room temperature lower than it would be if the circulation were stopped);
- reduces heat-loss to outdoors.

The bathing operation helps room comfort only when the outdoor temperature is very low. But the increase is usually insignificant, as explained on a previous page. When the operation does help room comfort, it does so at the cost of increasing overall heat-loss to outdoors.

CLAIM CONCERNING EFFECTIVE USE OF LUKE-WARM AIR

Some proponents claim that the double-envelope design has the special virtue of making use of below-70°F air to keep the house warm. They point out that, in typical solar-heated houses, heat must be collected and stored at above-70°F temperature; yet the lower the collect-and-store temperature, the greater the collection efficiency and the greater the storage efficiency.

They claim that, in a double-envelope house, luke-warm air (say air at about 60°F) can be put to effective use, with the consequence that collection and storage are efficient and especially simple.

What are the effective uses of the 60°F air?

Keeping the greenhouse from becoming too cold at night.

(This is certainly true, at least to a moderate extent.)

Thermally protecting the inner components of the exterior walls and roof.

(A previous page presents reasons for concluding that this benefit is extremely small -- negligible, usually.)

Preventing jets of cold air from entering the rooms.

(This benefit too has been shown to be extremely small.)

Thus the benefits seem to boil down to: keeping the greenhouse from becoming too cold at night.

"INTERCEPTION-OF-COLD-AIR" FALLACY

One designer has expressed the opinion that one of the main functions of the north wall airspace is to warm in-leaking air to at least 50°F or 60°F. The in-leaking air first enters this space, is warmed there, <u>then</u> may find its way into the rooms. Never does a jet of ice-cold air enter the room.

This argument is valid. But its relevance is low. Consider these facts:

There are no airspaces in the east or west walls. Thus the above-stated argument does not apply here. (Yet it is the west wall that bears the brunt of the prevailing wind!)

In an ordinary house there are few, if any, cold air jets entering via the roof. Leakage in the roof is usually outward, not inward. Thus the above-stated argument applies mainly just to the north wall.

In a modern, well-built house, there are many leaks, but most of these are in the basement, or crawl space, or at doors. There are few leaks in the north wall itself.

In a superinsulated house, with its extra-thick insulation and its overlapped and sealed vapor barriers, there are virtually no direct leaks.

REQUIRED SPEED OF AIRFLOW

A greenhouse with 400 sq. ft. of south-facing glass may take in about 80,000 Btu per hour. If, at noon, the greenhouse is already almost too hot, the designer would like all of this heat to be carried away by the gravity convective airflow. If a 20 F deg. temperature rise in this air is considered acceptable, then each cubic foot of air passing through will carry away this amount of heat: (0.077 lb/ft^3) (0.24 Btu/(lb. F deg.))(20 F deg. temp. rise) = 0.37 Btu. Thus to transport 80,000 Btu per hour requires an airflow of 80,000/0.37 = 216,000 ft^3/hr, or 3600 ft^3/min., or 60 ft^3/sec. If the cross sectional area of the airflow path in the crawl space is 32 ft. x 2 ft. = 64 sq. ft., the required linear rate of airflow here is about 1 ft./sec. Within the north wall space the required linear rate may be 2 or 3 ft./sec.

Little information as to actual rates is available.

Note: In making the above-presented calculation, I have not taken into account the amount of heat that is lost via the greenhouse glazed area. This loss is discussed below.

HEAT-LOSS VIA GREENHOUSE GLAZED AREA

The heat-loss, on a cold night, from greenhouse to the outdoors via the greenhouse glazing is large. Consider a 400 sq. ft., double-glazed greenhouse, and a night with 10°F outdoor temperature. If the temperature within the greenhouse were 60°F, the heat-loss during a 14-hour night would be:

$(400\ ft^2)\ (0.57\ Btu/(ft^2\ hr\ {}^oF))(50\ F\ deg.)(14\ hr/night) =$ 160,000 Btu/night.

This is about twice the loss through all of the walls of the house and the roof. In short, it is a very big loss.

On a sunny day also the heat-loss is very large. The day is shorter (6 or 8 hours), but the greenhouse temperature is much higher (say 80°F rather than 60°F). Outdoor temperature also is higher. Part of heat-loss from the greenhouse glass is made up by solar radiation absorbed by the glass.

This fact is of obvious importance: the hotter the air in the greenhouse during a sunny day, the greater the heat-loss from the greenhouse to outdoors; yet if much heat is to be carried (by the convective-loop airflow) to the crawl-space thermal mass, it is essential that the greenhouse be very hot. Thus there are two antithetical requirements:

Keep greenhouse temperature low, to minimize heat-loss here.

Keep greenhouse temperature high, otherwise little heat will reach the crawl space.

This somewhat parallels the paradox concerning heat-losses in roof and north wall: the losses must be kept low to avoid waste of much heat, yet the losses must be kept large in order to keep the speed of airflow high.

Again and again the question arises: Is there a successful middle ground? A successful compromise? Will the system adjust itself so as to provide such compromise? How often, or under what range of conditions, will a good compromise be achieved? Is it really practical to rely entirely on natural convection?

USE OF CONCRETE BLOCKS OR WATER FILLED-BOTTLES IN THE CRAWL SPACE

Instead of relying (for thermal storage) on crawl-space earth, one could use one of these alternatives:

> Fill a large fraction of the crawl space with one or two thousand 62-lb., 16 in. x 12 in. x 8 in., hollow-core concrete blocks, oriented so that the hollows are aligned to form scores of N-S passages for airflow. This scheme was invented and tried out successfully by Hank Huber in his New Ipswich, N.H., house.
>
> Obtain 1000 large glass bottles, e.g., discarded wine bottles, fill them with water, seal them hermetically, and install them very close below the ceiling of the crawl space. Thus they find themselves (on a sunny day) in the hottest stratum of the air moving along in the crawl space. The airflow is turbulent, thanks to the arrangement and spacing of the bottles. The aggregate surface area of the bottles is large. The crawl space will never get as cold as $32^{o}F$; thus there is no need to include antifreeze in the bottles.

IS THERE A BETTER LOCATION FOR AN ARRAY OF WATER-FILLED BOTTLES?

If an array of 1000 water-filled glass bottles is to be used, the designer might ask himself what is the optimum location, or optimum height of the location. For example, the bottles might be placed within the north wall system. If they were placed high up in the within-north-wall airspace, the speed of convective airflow on a sunny day would be speeded up -- because the process of extracting heat from the airstream in the north wall would be speeded and would occur higher up. The average density of the gas within this airspace would be greater, gravity would pull harder, and the flow would be faster. The north wall system might well be strong enough to support a large number of such bottles; or it could be strengthened for this purpose. Such a scheme might greatly increase the amount of heat stored on a sunny day (and decrease the amount lost out through the outer north wall).

However, if the array of bottles significantly blocked airflow within the north-wall space, it might do more harm than good. Also, it is barely conceivable that, at this location, the bottles might sometime become colder than $32^{o}F$; the water might freeze and burst the bottles.

Another drawback is that, situated high up in the within-north-wall airspace, the array of bottles would not give out heat efficiently during a cold night. Being high up, they would not produce a rapid airflow. For best nighttime performance they should be situated very low down, in the crawl space or bottom of the greenhouse.

An intermediate location might be best. Greater consideration should be given to sunny-day performance than cold-night performance inasmuch as the crucial sunny-day period is only 5 hours long, whereas the night is about 14 hours long.

WHY NOT BEEF UP THE WITHIN-GREENHOUSE THERMAL MASS?

In many greenhouse-equipped solar houses a large amount of thermal mass, with large area, is incorporated in the greenhouse. My impression is that such scheme, although preempting some space and causing some inconvenience, can be very effective in keeping the greenhouse warm at night and preventing it from becoming too hot during sunny days.

Should this approach be used in the houses we are now discussing? Probably there are many pros and cons.

WHY NOT USE DUCT AND BLOWER?

Many solar architects have told me that they believe that the simple, cost-effective way to transfer hot air from the upper part of the greenhouse to a thermal storage system is by a duct and blower. The necessary hardware is readily available, costs are low, and the engineering know-how is well developed. Such a system takes up little space. There is great flexibility of layout. There are easy ways of controlling such a system: heat can be transferred at the rate you choose, when you choose, to the location you choose; it can be delivered to the bottom (rather than the top) of the thermal mass.

Two limitations are (a) the noise made by the blower and (b) the dependence on the electric supply -- if it fails the solar heating system fails. However, the noise is small and can be made smaller. And if the thermal mass is at some low-down location, some flow of heat upward from this mass will occur even when the blower is not running. The house is well insulated and has long carrythrough; electrical-supply interruptions lasting only a few hours would cause little trouble.

Perhaps in a few years it will be feasible to install an array of photovoltaic cells that, on sunny days, will provide enough power to run the blower. This would eliminate the dependence on the electric utility company.

WHAT IS THE OPTIMUM HEIGHT OF CRAWL SPACE?

Speakers at the July 21, 1979, double-envelope-design conference in Boscawen, N.H., seemed unable to come up with a simple answer to this question. There is this contradiction:

> The height must be great if the pneumatic conductance is to be great. The conductance is <u>proportional</u> (I think) to the crawl-space height if the flow is turbulent, and is proportional to the second or third power of the height if the flow is laminar.
>
> The height must be small because of the tendency of the warmest air to hug the ceiling of the crawl space. If, during a sunny daytime, much heat is to be imparted to the earth, the ceiling of the crawl space must be close to the earth. In other words, the crawl space must be thin.

There is the added complication that if people are really to be able to crawl in this space, the height should be at least 2 ft. and preferably 3.

WHERE ARE VAPOR BARRIERS NEEDED?

My impression is that there is agreement that vapor barriers are needed on the warm sides of the outer north wall and outer roof.

Whether a barrier is needed on the cold side of the inner wall and the inner roof seems unclear.

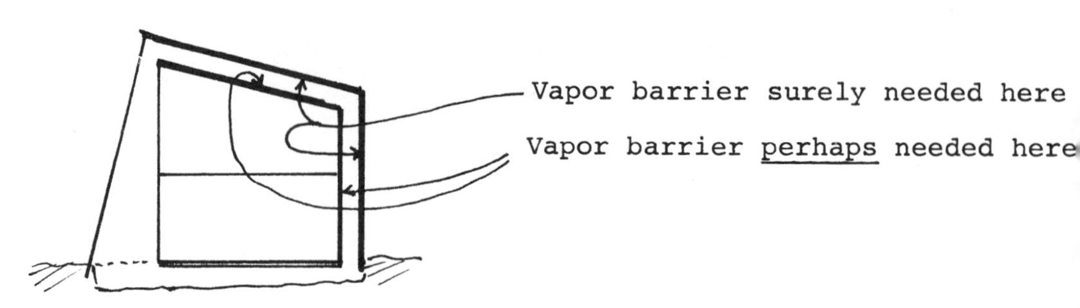

CONDENSATION OF WATER VAPOR IN CRAWL-SPACE EARTH: DOES THIS HELP?

Speakers at the Boscawen conference seemed to suspect that much of the heat transferred, on sunny days, to the crawl-space earth is transferred by virtue of condensation of water vapor in this earth. They reasoned thus: When the greenhouse is irradiated, and warms up, much moisture here evaporates; then the moist warm air circulates around the grand loop and eventually comes in contact with the crawl-space earth, and much condensation occurs here. The condensation process delivers much heat.

I agree that, in principle, much heat could be imparted to the earth by this mechanism.

But I see this discouraging consequence: during the night, approximately an equal quantity of water will evaporate -- thus "using up" most of the heat which the sunny-day process delivered here. And, over the long term, the amount of evaporation will roughly equal the amount of condensation. Accordingly, most of the heat delivered to the crawl-space earth by the condensation mechanism is not available for any useful purpose.

(The within-crawl-space condensation that occurs during a sunny day may be useful in disposing of unwanted moisture during such day and re-supplying moisture at night. That is, it may be useful as a humidity regulator. But it would seem not to help keep the rooms or greenhouse warm at night.)

GREAT TOLERANCE AS REGARDS OPERATING CONDITIONS

Reading the summary of the Boscawen Conference, one gets the impression that the tolerances on operating conditions are very great; that is, wide variations in design and control make little or no difference.

Whether this great tolerance is reassuring and to be welcomed, or whether it casts doubt on the main principle (or designers' understanding of the principle), I do not know.

What are some of the tolerances? One gathers from the report in question that performance hardly changes:

whether the air descending at the north descends in the within-north-wall airspace or descends via the north rooms themselves,

whether the within-north-wall airspace is especially wide or especially thin, or is fully clear or is somewhat obstructed,

whether air actually is circulating, or is not circulating, in the crawl space,

whether the doors and windows between greenhouse and south rooms are open or shut,

when different designs of ceiling or attic airspaces are used,

whether a wood stove is, or is not, in use.

MEANS OF REDUCING FIRE HAZARD

If, in the grand loop, there are no dampers that will come into play quickly on the outbreak of fire, the fire may spread rapidly, according to an article "Conservationist's Dream...Firefighter's Nightmare" by Bill Pfaehler, in the Maryland Fire & Rescue Bulletin of early 1979 (?). It is claimed that the fire might spread very rapidly and soon involve most of the grand loop, with the result that the occupants of the house might find themselves inside an oven with many escape routes cut off. Designers are urged to plan carefully to prevent such catastrophe. (I do not know what particular kind of grand loop Pfaehler had in mind.)

Some speakers at the July 21, 1979, Conference at Boscawen, N.H., indicated that the fire threat can be reduced satisfactorily by the installation of within-airspace dampers that drop into place automatically when the heat reaches certain fusible links.

Other speakers suggested that it would be necessary to install gypsum boards on both faces of the within-wall airspace.

The use of within-airspace smoke detectors, fire extinguishers, or sprinklers was proposed. But who would test these devices periodically? Would they really work when needed? Would water in the sprinkler-system pipes freeze and burst in a long cold cloudy period in winter in which all of the occupants were away?

Could it be that the fire hazard is overstressed? If there is no furnace or stove in the crawl space -- indeed virtually nothing at all there -- how could fire originate there? (I have heard a rumor that a double-envelope house in California burned down. I do not know the cause of the fire.)

Recommendations from California: In May 1980 the Resource and Energy Committee of the San Diego Chapter of the American Institute of Architects submitted some formal recommendations to State of California officials concerning:

emergency exit routes from bedrooms of double-envelope houses,

fire resistive construction of such houses, or use of smoke alarms and sprinkler systems,

firestopping the within-wall airspaces by means of automatically activated dampers.

The detailed recommendations were distributed by J.P. Brown (of Donald J. Reeves and Associates, 835 5'th Ave., San Digeo, CA 92101), Chairman of the above-mentioned committee. Committee address: 233 A St., San Diego, CA 92101.

HAZARDS ASSOCIATED WITH SLOTS IN GREENHOUSE FLOOR

In many double-envelope houses the greenhouse floor includes a large slotted area: a large grille that permits flow of air. Often, there is a large space below, where adults could be working, talking or sleeping, or children could be playing or flammable objects could be stored. Suppose a person in the greenhouse drops a burning match or cigarette through a slot, or while spraying allows some poisonous liquid to fall through the slots, or while sweeping sweeps some sand through the slots. Could not this create the danger of fire, or pose a threat to persons' eyes or hair or skin? Might not the person in the greenhouse be unaware that there were persons below, and unaware of the start of fire or injury to persons there?

In general, is it not a bad idea to have an indoor area with a floor that consists of a grille and, below, another region that is not visible from above yet may contain people or may contain flammable objects? Should all persons concerned be warned not to place flammable objects below the greenhouse, and, in addition, be warned not to look upward when in the space below the greenhouse?

TWO CLASSES OF INTEGRAL GREENHOUSES

A greenhouse that is integral with the south side of a house may be non-intervening or intervening with respect to house and south lawn. If it is non-intervening, persons in the house can walk directly onto the lawn without having to pass through the greenhouse. If it is intervening, persons in the house must pass through the greenhouse in order to walk to the south lawn; they may have to operate two sets of doors (at north and south sides of greenhouse) and may have to detour around gardening tools, hose, boxes of seedling, etc. If the doors are of sliding type, strength is needed to slide them; and, after persons have passed through, the doors may have to be pushed shut again. (Small children may be unable to slide the doors.) (Of course, at some times of year, or some times of day, some of the doors in question may remain open continually.)

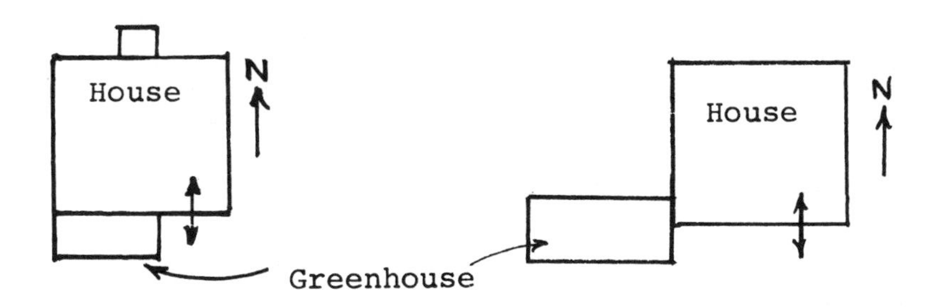

Examples of non-intervening greenhouse that adjoins a house. To walk from house to south lawn is simplicity itself: see double-pointed arrow. The door may be of simple swinging type; if it is, it is likely to be of self-closing type. The house shown at the left is patterned on a house built in Bridgeton, MO, and designed by J.A. Ray et al of Ener-Tech, Inc.

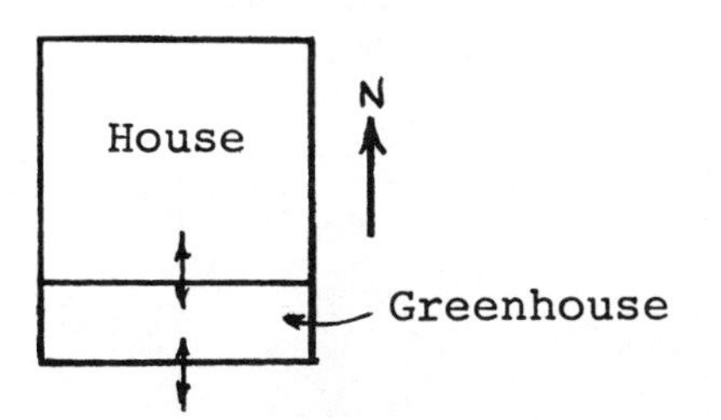

Here the greenhouse is of intervening type. To walk from living room to south lawn, persons must operate two doors and these are likely to be of sliding type -- difficult to operate and not self-closing.

Sacrifice of Rapport With The South

If the greenhouse is of intervening type, all of the south windows of the rooms open onto the greenhouse only. There are no windows (of rooms) that open to the great outdoor south. To achieve close rapport with the greenhouse, rapport with the outdoor fragrance and outdoor scene (trees, hills, clouds to the south) is sacrificed. I feel unhappy about this sacrifice -- this partial divorce from the scene to the south. I tend to favor greenhouses that adjoin but do not intervene.

THERMAL PERFORMANCE OF DOUBLE-ENVELOPE HOUSES: SOME PRELIMINARY DATA

Several groups are now monitoring and analyzing the thermal performance of double-envelope houses.

I know of no final, or accurate, results.

The preliminary results are so rough and tentative that I do not consider it helpful to present results from specific houses.

From what I learned of results available early in 1980, I have the impression that:

The houses are generally comfortable.

The amount of auxiliary heat needed in winter is small.

On a sunny day some hot air does indeed circulate, purely by gravity-convective flow, around the big convective loop. In particular, some slightly warm air does arrive at the crawl space and impart some heat to the crawl space earth. But the airflow speeds are low, seldom exceeding a few inches/sec.; usually the speed is less than ½ ft./sec. Even if the temperature of the air near the top of the greenhouse is high (of the order of 100°F), the temperature of the air as it reaches the bottom of the within-north-wall space is not very high; sometimes (if outdoor temperature is very low) it may not be much higher than 50°F.

Some -- but not much -- heat is stored in the crawl-space storage material. But the range of temperature change here is small: of the order of a few F degrees, in a 24-hr. period involving bright sun and dark night, with the temperature sensors located at various depths (from 0 to a few inches) in the storage material; that is, the amount of energy stored during sunny periods and given out at night is small.

Recent data: By Sept. 1, 1980, interesting data on performance of several double-envelope houses had become available. Included were data on: Burns House, Mastin House, New Ipswich House, Kootenay House #1. However, few general, or decisive, results were available. A few of the more interesting results are shown on pages 102, 110, 121, and 125.

NOTES ON THE PHYSICS OF AIRFLOW IN THE CONVECTIVE LOOP ON A SUNNY DAY

Consider the simplified convective loop shown in the sketch at the right. Assume that solar energy is delivered just at one location P, and assume that energy is extracted from the flowing air just at the location Q at the top of the north leg and in the crawl space.

Then the laws of physics dictate that air in the loop will circulate spontaneously, i.e., by gravity convective flow, and the sense of circulation will be clockwise. Why so? Because the air in the north leg is cooler than the air in the south leg, and is denser, and accordingly gravity pulls harder on it than on the south-leg air.

The flow rate may be low because of the pneumatic resistance, i.e., friction.

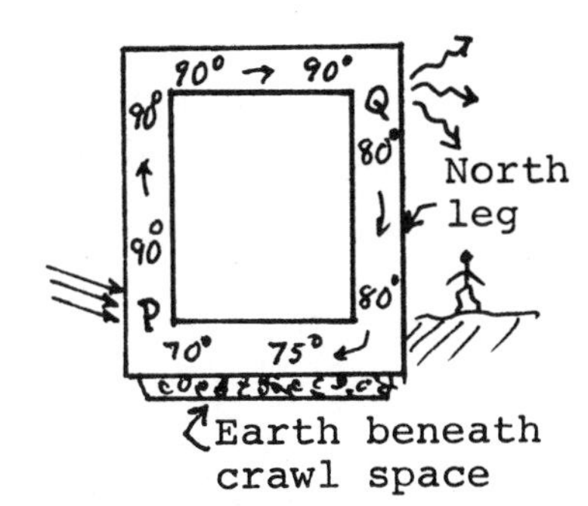

Vertical cross section of simplified convection loop, looking west

Suppose that the loop extends as far below P as it extends above it. Suppose that the amount of heat extracted at Q slightly exceeds the amount extracted in the crawl space. Will there be a flow? Will the sense of circulation be clockwise?

Yes, because the average temperature in the north leg is slightly less than that in the south leg. The density in the north leg will be slightly greater than the average density in the south leg, and gravity will pull harder on the north-leg air.

The flow rate may be very low, because (1) the difference in average densities is small and (2) there is friction.

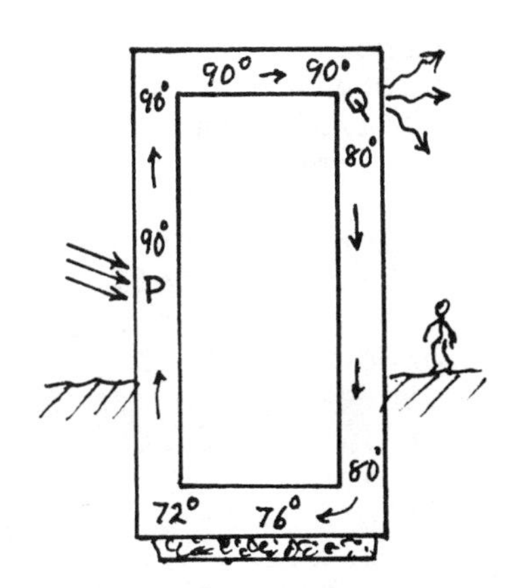

Rule: The farther the crawl space is below point P, the greater this ratio must be:

$$\frac{\text{heat extracted at Q}}{\text{heat extracted in crawl space}}$$

-- i.e., the smaller is the fraction of the heat than can be delivered to the crawl space.

It is clear also that:

If some of the absorption of solar energy occurs higher up in the south leg, the situation becomes less favorable. Why? Because the average temperature in the south leg is then less high, and the density is less low -- the reverse of what one would wish.

If some of the extraction of heat from air in the north leg occurs lower down in that leg, the situation becomes less favorable. Why? Because the average temperature in the north leg is then less low, and the density is less high -- the reverse of what one would wish.

In fact, in a double-envelope house, the energy absorbed in the greenhouse is absorbed throughout a range of heights. Much is absorbed near the base (floor), but some is absorbed higher up.

And, in the north wall, heat extraction occurs at all locations, from top to bottom. Some extraction occurs in the roof, and this in a sense helps -- provides a kind of head-start to the cooling down within the north leg.

In summary, merely from considerations of gravitational forces, one concludes that, at most, only a modest fraction of the solar energy delivered to the greenhouse can reach the crawl space and be absorbed there.

And to arrange for even a modest fraction to be absorbed there, the designer must make sure that much energy is extracted from the circulating air at some high-up location, such as upper part of north wall, or roof. How does he arrange this? By having only a moderate amount of insulation between the within-north-wall airspace and the outdoors. That is, by deliberately throwing away heat.

Friction: Another important controlling factor is friction. The walls that define the airspaces, and also the projecting joists, cross-ties, etc., slow the airflow. An even greater slowing is produced by the corners, or "bends", in the path -- which tend to decrease the momentum of the air. Turbulence occurs.

What fraction of the heat produced by the absorbed solar radiation may be expected to reach, and travel along, the crawl space of a double-envelope house? I don't know. I guess that the fraction is small: 1/8? 1/16? 1/24?

The crucial question: what fraction of the heat produced by the absorbed solar radiation will penetrate into the thermal mass beneath the crawl space? I don't know. I guess that the fraction is small or very small: 1/10? 1/20? 1/30?

Let us hope that, soon, quantitative and reliable answers to this question will be available.

Question as to where most of the absorbed heat does go: If very little heat finds its way into the crawl-space thermal mass, where does the majority of the heat go? My guess is that a large fraction goes back out through the greenhouse glazing. The R-value of this glazing is only about 1.8, very small compared to the R-19 (say) value of the insulation between north-wall airspace and outdoors. Having such a large area, and such small R-value, the greenhouse must lose an enormous amount of heat to the outdoors. Thus there is a paradox here too: to deliver much heat to the crawl space, the driving temperature (temperature of air in the greenhouse) must be high; but if it is indeed high, the heat loss via the greenhouse glazing is enormous.

UNDERGROUND PIPES FOR SUMMER COOLING

Several double-envelope houses are equipped with large-diameter (1 or 2 ft. diameter) pipes buried about 6 ft. deep in the earth; the pipes may be 50 ft. long, with one end opening into the crawl space or basement and the other end open to outdoor air. Both ends are screened to keep out flies, mice, etc. When hot air in the house is vented via openings in the gables, outdoor air is drawn into the house via these pipes. Being in deep-down cool earth, the pipes cool the incoming air.

Although the system is certainly successful in principle, it has some limitations, listed in a survey article by Douglas B. Nordham, of SERI, on "Design Procedure for Underground Air Cooling Pipes Based on Computer Models", Proceedings of the 1979 Conference in Kansas City, p. 525. He concludes that, although the ground has great capacity for storing heat or for keeping cool, the interfacing of the air (in the pipes) to the ground is difficult. Greater transfer area would help. He suggests using several smaller-diameter pipes (arranged about five diameters apart) rather than a single large-diameter pipe.

Possible alternative: Use an extra large, or extra-deep basement. Its walls will remain somewhat cool throughout the summer. Arrange for introducing air into the rooms via the basement, and arrange for outdoor air to enter the house via the basement. The basement has many uses; only a very small fraction of its cost is to be attributed to its role in supplying cool air to the rooms.

OVERALL CRITERIA TO BE USED IN JUDGING A DOUBLE-ENVELOPE HOUSE

In Chapter 6 I explain the difficulty of finding objective criteria for use in judging the success of a superinsulated house. In judging double-envelope houses, also, such difficulty exists. The amount of auxiliary heat needed is small -- almost trivial; exact evaluation of such need is of little use.

The main criteria are those relating to comfort, pleasure (from greenhouse, pleasant smell, fresh home-grown vegetables, independence of oil dealers), and other hard-to-measure satisfactions.

Cost of construction is, of course, a very important criterion. Durability is another.

Chapter 11

THE NORWEGIAN HOUSE AT AAS

This house, while having many features of the house-types I call superinsulated and double-envelope, differs in important ways from those houses. The Aas house has a double envelope, but (1) the airspace between the two envelopes is much too thin to encourage gravity-convective airflow, (2) the main airflow is forced by blowers, (3) most of the solar energy received is received by a sloping collector on the roof. Also, the greenhouse is not a part of the main convective loop.

For these reasons, this house is described in a separate chapter -- this chapter.

* * * *

NORWAY

<u>Aas, Norway</u> (Suburb of Oslo)

<u>Aas House</u>: Danskerudveien 12, N-1432 Aas, Norway. 60°N. A 8200-F-degree-day location. Completed in 1978.

<u>Architect and engineers</u>: Johannes Gunnarshaug, Fritjof Salvesen, and Anne G. Hestnes. Sponsor: SINTEF (Society for Technical and Scientific Research) of the Norwegian Technical University, Trondheim. Occupant: Nils Skaaser.

<u>Building</u>: Two-story building the plan-view dimensions of which are 11 m x 7.5 m (36 ft. x 25 ft.). The floor area of the space that is kept warm night and day is 100 m^2 (1080 sq. ft.). The attic includes considerable space that is not fully heated. There is a small integral greenhouse at center of south side; it is 6.6 m x 2 m (22 ft. x 6½ ft.). On the sloping south roof there is a large air-type collector. There is a crawl sapce. No basement, no garage. Nearly all of the outer walls, including east and west walls as well as north and south walls, are double, with a thin airspace between.

Aas House, perspective view.

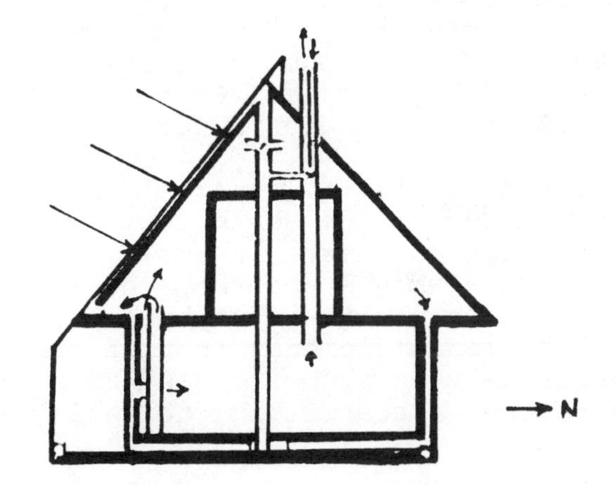

Vertical cross section, looking west

Wall: A typical exterior wall system includes an outer wall, inner wall, and intervening airspace 4 cm. (1.6 in.) thick. The outer wall (likewise the inner wall) consists of a prefabricated (factory-assembled) sandwich, or plate. The plate is 7 cm. (3 in.) thick, 2.4 m (8 ft.) high, and may be as long as 13 m (43 ft.). It includes, on both faces, a 1-cm.-thick plaster board. The core, which is 5 cm. (2 in.) thick, is of urethane foam. Such plates are cheap: about $\$1/ft^2$, delivered. When the plates are installed, the joints are carefully sealed with tape. The urethane foam serves as vapor barrier.

Windows: The area of windows between living space and outdoors is 8½ m^2 (90 ft^2). This is made up of: S, 10 ft^2; W, 40 ft^2; N, 10 ft^2; E, 30 ft^2. There are additional windows between living rooms and greenhouse. The windows are quadruple glazed. (Initially there were two glazings in the inner wall and one in the outer wall; but condensation occurred on this latter glazing and accordingly, in the summer of 1979, an additional glazing layer was installed.) In each wall -- inner and outer -- there are two glazings, making a total of four. The openings (for windows) in the big wall plates are made on-site; they are made with a saw. No thermal shutters or shades are used.

First-story floor: This includes a 5-cm. (2-in) layer of mineral wool which rests on a 15-cm. (6-in.) layer of foam concrete.

Crawl space: Below the first-story floor there is a 40-cm. (16-in.) crawl space. Below it is a gravel bed, described below.

Eaves: At north and south there are eaves that extend 1 m (3 ft.) beyond the walls.

Vapor barrier: The urethane foam cores of the wall panels serve as vapor barriers.

Humidity: After the glazing of the windows was increased to quadruple, no significant humidity problems have been encountered.

Fresh air: This is introduced via airspaces within the walls. Here stored heat from the gravel bed and other massive components warms the fresh air. Thus, to be effective, solar energy that has been stored needs only to have the form of heat at intermediate temperature (below-room-air-temperature); thus the storage system itself does not have to be very hot, and accordingly solar energy may be collected at very modest temperature -- with the consequence that collection efficiency remains reasonably high even on mid-winter days that are fairly cloudy. Note: Tests have shown the concentration of radon in room air to be higher than had been hoped. Perhaps the gravel is responsible (and should not have been used?)

Solar collector: The 45 m^2 (480 ft^2) air-type, active, collector is mounted on the south roof which slopes about 45°. The collector was built on-site. It employs single glazing, of glass, and has a black aluminum absorber sheet. Air circulates behind this sheet; a blower circulates the heated air through the storage system. Collection efficiency is high because the storage system is only moderately warm, being used mainly to heat the (below-room-temperature) space within the wall systems.

Storage system: Short-term storage is provided by a bed of 5-cm (2-in.) dia. gravel. The 13 m^3 (450 ft^3) of gravel is in the form of a horizontal bed 15 cm.(6 in.) thick. The gravel is confined between two sheets of plastic. The lower sheet rests on the earth below the crawl space. The earth itself serves as long-term storage.

Auxiliary heat: Electric furnace.

Domestic hot water: Preheated in pipes buried in storage-system gravel.

Convective loops; airflow: As explained above, all four walls are of double-envelope, or double-shell, type. Between the two wall elements (outer and inner) there is a 3.5 cm. (1½ in.) air-space. This space is not thick enough to encourage or sustain gravity-convective airflow. Blowers are used to maintain the flow. There are three blowers. Also there are several vertical ducts, and there are dampers or valves or vents. In general, the hot air in the collector on a sunny day is not delivered to the rooms; rather it is delivered to the storage system (gravel) in the crawl space. The hot air from the greenhouse is put to very little use; it may flow into the south rooms if the pertinent doors (between rooms and greenhouse) are open; but it does not flow into the within-wall airspace or into the storage system. (The purpose of keeping collector air and greenhouse air separate was to avoid moisture problems associated with very-moist green-house and avoid algae problems. Whether the reasoning was sound is uncertain, and in subsequent houses a different strategy is being followed.)

There are no within-airspace fire-stop dampers and fusible links: These are deemed unnecessary, inasmuch as the surfaces that define the airspace are sheets of (non-flammable) plasterboard.

Calculated rate of heat-loss: By house (not including greenhouse or collector) when the air is changed 0.5 times per hour: 150 W/°C; i.e., 284 (Btu/hr)/°F.

Thermal performance: Amount of auxiliary heat used per winter: 5800 kWh. (Note: if electric power were to cost 5¢/kWh, the total cost here is about $300.) It is said that, in this same location, a house built according to typical modern Norwegian standards would require 2 or 3 times as much auxiliary heat. Even in very cold weather the present house requires only 2½ kW. Early in 1980 a detailed analysis of the performance of the building was being completed.

References:

Personal communication from Johs. Gunnarshaug.

Article in "Proceedings of Conference on Solar Energy North of the 50th Parallel, Stockholm, Sweden, May 1978".

J. Gunnarshaug, F. Salvesen, and A.G. Hestnes, "Ressursvennlige Boligformer Solvarmehus, Danskrud, Ås Kommune; Del II", Report STF62 F77009. Published by SINTEF (Selskapet for Industriell Og Teknisk Forskning Ved Norges Tekniske Høgskole), Norwegian Technical University, Trondheim. About 50 pages, with about 40 detailed drawings. 1977.

Many other articles, mostly in Norwegian, by Johs. Gunnarshaug et al of SINTEF.

POSTSCRIPT CONCERNING NORWEGIAN HOUSE AT AAS

Any effort to make a close evaluation of this house may be moot, or at least premature, inasmuch as the designers have recently changed their minds as to some of the design elements. I understand that, in future houses, they may:

abandon the use of a gravel-type storage system; there is concern that the gravel may be emitting some radon,

cease practically isolating the greenhouse air from the collector airflow paths and from the storage system.

I question whether the large collector on the sloping roof is cost-effective. And whether it is needed.

Is the 1½-in. airspace between inner and outer walls cost-effective? Its cost is very small indeed: it comes "free" once one has bought and installed the inner and outer wall panels. Also it preempts very little space. It does not contribute to any fire hazard. Yet does it perform any valuable service? Would the house keep just as warm if this space were filled with fiberglass? Or if, in addition, a well-sealed polyethylene vapor barrier were installed on the inner wall?

Is the overall system somewhat complex, with its several blowers, several ducts, and special dampers, valves, vents? --With its greenhouse, collector, crawl space, gravel bed?

Is the enclosed volume that is <u>not</u> heated day and night excessively large?

* * * *

Although the house is undoubtedly highly successful and a well-engineered example of pioneering construction, may it not be superseded by other designs? Is it perhaps already superseded by superinsulated houses such as are described in Chapters 3 and 4?

Chapter 12
TENTATIVE COMPARISON OF SUPERINSULATED HOUSES AND DOUBLE-ENVELOPE HOUSES

INTRODUCTION

These two kinds of houses -- superinsulated and double-envelope -- are competing. Each is new, stimulating, attractive, and saves energy. Each has its loyal group of boosters.

Although I feel across-the-board enthusiastic about superinsulated houses but have doubts concerning some important features of double-envelope houses, I am sure it is too early to make a final judgment. In any event, judgments should be made by architects, not physicists.

Perhaps in a year or two there will be enough firm information on performance and cost to permit architects to make reliable judgments.

PRELIMINARY ATTEMPT AT A COMPARISON

The superinsulated house seems to me to be an already-demonstrated success. Actual houses have been built and lived in, and perform excellently. Also, groups working with computers have analyzed the designs in detail, and their computations bear out what is actually happening. The actual performance agrees closely with the calculations.

Also, the principles involved -- the mechanisms, and individual kinds of behavior -- are well understood. The subject is under good control.

There are no major problems remaining. True, humidity may sometimes be high; but this can be taken care of by opening windows or employing dehumidifiers or air-to-air heat exchangers. True, the houses built to date are not very exciting looking and not much "fun". But next year's designs may overcome these limitations -- and, anyway, in an era of energy crisis and cold homes, we cannot afford to spend much time planning for excitement and fun.

The double-envelope houses, contrariwise, appear to be in an early stage of development. Designers are uncertain as to which features are crucial and which add little (or add nothing); present designs may be far from optimum. There are few (if any) in-depth, quantitative, computer studies showing just how each component performs and just what the optimum shapes, dimensions, and controls are. Present designs are exciting and fun, but they have the hallmarks of experiment, of cut-and-try.

CATEGORY-BY-CATEGORY COMPARISON

The following tabulation summarizes many of the differences between the two types of design.

Category	Superinsulated House	Double-Envelope House
Greenhouse	Has none. A greenhouse would lose much heat at night and thus greatly impair performance.	Has integral greenhouse -- an important and much-liked feature.
Requirements as to full exposure to sun in winter	Moderate exposure may suffice. Only a moderate amount of solar energy is needed.	Large. Greenhouse glazed area must be great and much solar radiation must be received.
Comfort in winter	Excellent.	Excellent. But basement ma become cold on cold nights inasmuch as it serves as a catch-basin for cold air from the greenhouse.
Rate of air change in winter	May sometimes be too low or almost too low. Use of air-to-air heat-exchanger may be necessary.	Fully adequate.
Access from south rooms to south lawn	Direct access is easily provided.	Indirect. The greenhouse intervenes.
Access from south rooms to south air and sky	Direct. Merely open south windows.	Indirect. Windows of south rooms open to greenhouse only.
Requirement as to large thermal mass	No requirement. There are no sudden large energy inputs or sudden large loads.	Urgently required because c sudden large solar energy input in 6-hr. period at mid-day and large heat-loss via greenhouse glazing at night.
Amount of auxiliary heat needed	None. Or almost none.	Small.
Wood stove or equivalent	Not needed.	Needed.

Category	Superinsulated House	Double-Envelope House
Amount of enclosed volume that is not kept warm at night	Practically none. Only the attic space is cold.	Large. Includes greenhouse, within-roof space, within-north-wall space, crawl space.
Spaces accessible to animals and children but not to adults	None.	Several. See item directly above.
Spaces requiring fire-stop dampers	None.	Within-northwall space (?)
Cost of thermal shades -- if they are needed	Small. Windows are small. (Perhaps shades are not needed.)	Large. Greenhouse glazed area is large. (Perhaps shades are not needed.)
Large vents in gables (for summer use)	Not needed. There is no big heat load to dissipate.	Greatly needed to vent large amount of hot air in (unshaded) greenhouse.
Cooling requirement in summer	Little cooling needed. Outdoor air cools rooms at night. On sunny day house is kept closed and eaves block direct radiation.	Solar heating of (unshaded) greenhouse is very large, necessitating rapid air change. The incoming air may itself be hot unless it is drawn in via a long, large-diameter, buried pipe.
Size of air conditioner needed on hottest days	One small conditioner suffices.	Very large size would be needed (except to cool just the north rooms) because the greenhouse is large and unshaded.
Privacy	Excellent. Windows are small and sills are high above ground.	May be a problem inasmuch as greenhouse glazed area is large, glazing extends down close to ground. <u>Inner</u> south wall contains much glass, some of which extends down to floor level.
Attraction to vandals	Small. See item just above.	May be appreciable because of the large area of glass extending down close to ground and because passers-by can see into the house fairly easily.
Durability and ease of maintenance	Can be excellent.	Can be excellent.
Incremental cost of superinsulation or greenhouse-and-convective-loop	Said to be small: about 0 to $2000. --Small because saving re furnace etc. almost cancels cost of extra material and labor.	No firm data. My guess: $2000 to $10,000. Or more if one includes cost of volume preempted by the special airspaces.

Appendix 1

ORGANIZATIONS AND INDIVIDUALS INVOLVED

Acorn Structures, Inc., Box 250, Concord, MA 01742
Kelley, Mark E., III. Double-envelope house design.

Adirondack Alternative Energy, Edinburg, NY 12134.
Brownell, Bruce R. President. Design of heavily insulated houses.

Alternative Technology Associates, PO Box 503, Davis, CA 95616.
Maeda, Bruce. Double-envelope house design.

Brookhaven National Laboratory, Upton, NY 11973.
Jones, Ralph F.
Krajewski, Paul.

Community Builders, Shaker Rd., Canterbury, NH 03224. (603) 783-4743.
Booth, Don. Head. Builder of double-envelope houses.

Domestic Technology Institute, PO Box 2043, Evergreen, CO 80434.
Lillywhite, Malcolm. Double-envelope house design.

Dovetail Press, Ltd., PO Box 1496, Boulder, CO 80306.
Jordan, Roger. Double-envelope houses.

Ekose'a, 573 Mission St., San Francisco, CA 94105.
Butler, Lee P. Double-envelope house pioneer; designer; architect.

Enercon Consultants Ltd., 3813 Regina Ave., Regina, Saskatchewan, Canada S4S 0H8. Affiliate of Enercon Building Corp.
Lange, Leland
Rogoza, Dennis D. President. Has designed superinsulated houses.

Ener-Tech, Inc., 1924 Burlewood Dr., St. Louis, MO 63141.
Ray, J.A., Vice-President. Designed double-envelope house built in Bridgeton, MO.

Hart Development Corp., 5808 River Dr., Lorton, VA 22079. (703) 339-5590.
Also (703) 569-1947. Involved in building about 44 superinsulated houses.

Harvard University, Cambridge, MA 02138.
Shurcliff, William A. Physics Dept. Home address: 19 Appleton St., Cambridge, MA 02138.

Kern and Spivack, 6940 Hwy 73, Evergreen, CO 80439.
Spivack, Joanne D. Principal. Designer of double-envelope houses.

Kirkwood Community College, Passive Solar House Construction and Education Program, 6301 Kirkwood Blvd., Cedar Rapids, IA 52406.
Bean, Larry L. Program Manager. Superinsulated house program.
Driscoll, Ralph. Carpentry instruction. Superinsulated house constructior

Mid-American Solar Energy Complex, Alpha Business Center, 8140 26 Ave. S., Minneapolis, MN 55420.
Scott, Michael G. Superinsulated houses.
Robinson, David A. Design of superinsulated houses.
Pogany, David. Manager, Passive Initiative.

National Research Council of Canada, Division of Building Research, Prarie Regional Station, Saskatoon, Saskatchewan, Canada S7N OW9.
Dumont, Robert S. Superinsulated house design. Monitoring of Saskatchewar Conservation House.

New Alchemy Institute, 231 Hatchville Rd., East Falmouth, MA 02536.
Baldwin, J. Experimented with superinsulated dome.

Norwegian Technical University, Trondheim, Norway.
Selskapet for Industriell og Teknisk Forskning (SINTEF 62), i.e., Society for Technical and Scientific Research.
Gunnarshaug, Johs. Design and monitoring of double-envelope houses of special type.
Salvesen, Fritjof. Design of double-envelope houses.

One Step Ahead Energy Systems, Ltd., 301 Vernon St., Nelson, BC Canada V1L 4E
Early, Allan M. Double-envelope house design.
Rodgers, Russell. Double-envelope house performance.

(Popular Science Monthly: see Times-Mirror Magazines)

Positive Technologies Corp., PO Box 2356, Olympic Valley, CA 95730.
Smith, Thomas. Head. Owns double-envelope house. Co-author of book.

Princeton University, Princeton, NJ 08544.
Center for Energy and Environmental Studies, Engineering Quadrangle.
Dutt, Dr. Gautam S. Analysis of superinsulated houses.

Project 2020, Box 80707, College, AK 99708.
McGrath, Ed. Wrote book on superinsulated houses.

Saskatchewan Department of Mineral Resources, 1404 Toronto-Dominion Bank Building, Regina, Saskatchewan, Canada S4P 3P5.
Eyre, David. Engineer of Saskatchewan Energy Conserving House Project.
Rogoza, Dennis D. Was associated with this department until August 1978.

Smith, (Robert O.) and Associates, 55 Chester St., Newton, MA 02161.
Smith, Robert O. Performance monitoring of superinsulated house.

Solar Clime Designs, Box 9955, Stanford, CA 94305.
Berk, Jim. Head. Publishes periodical Convection Loop.

SRI International, 333 Ravenswood Ave., Menlo Park, CA 94025. Formerly Stanford Research Institute.
Gerlach, Kenneth A. Double-envelope house design.

Super Insulated Homes, PO Box 73876, Fairbanks, AK 99707.
Roggasch, Bob. Superinsulated houses.

Thacher & Thompson, 215 Oregon St., Santa Cruz, CA 95060.
Rahders, Richard P. Architect. Designer of double-envelope houses. Wrote book. The company, a general contractor, has built several such houses.

Times-Mirror Magazines, 380 Madison Ave., New York, NY 10017.
Publishes Popular Science Monthly.
Dans, Ron. Author of articles on double-envelope houses. Home: 123 W. 93 St., Apt. 7-D, New York, NY 10025. (212) 662-6789.

Total Environmental Action, Inc., Church Hill, Harrisville, NH 03450.
Pietz, Paul. Designed superinsulated house: Brookhaven House.

(United States of America: see Brookhaven National Laboratory)

University of California, Lawrence Berkeley Lab., Berkeley, CA 94720.
Roseme, G.D. Air-to-air heat-exchangers.
Rosenfeld, A.H. Energy conserving buildings.

University of Illinois, Urbana, IL 61801.
Small Homes Council and Building Research Council, 1 East St. Mary's Rd., Champaign, IL 61820.
Harris, Prof. Warren S. Died in 1979. Superinsulation.
Jones, Prof. Rudard (sic) A. Superinsulation.
McCulley, Michael T. Designer and builder of superinsulated house. Office: (217) 333-1911. Home: 4003 Farmington Dr., Lincolnshire Fields West, Champaign, IL; (217) 351-9113.
Konzo, Prof. Seichi. Superinsulation.
Shick, Prof. Wayne L. Home: 1024 W. Charles St., Champaign, IL 61820. Pioneer in superinsulation and key member of team designing the Lo-Cal superinsulated house.

University of Nebraska, Passive Solar Research Group, 60th and Dodge Sts., Omaha, NE 68182.
Chen, Dr. Bing. (402) 554-2769. Monitoring performance of double-envelope house.

University of Saskatchewan, Saskatoon, Saskatchewan, Canada S7N 0W0.
Department of Mechanical Engineering
Besant, Robert W. Superinsulated houses.
Schoenau, Greg J. Superinsulated houses.
Van Ee, Dick. Design of air-to-air heat-exchanger.
Dumont, Robert S., Div. of Bldg. Research, 110 Gymnasium Rd., S7N 0W9. Superinsulated houses.

Vista Homes, PO Box 95, E. Pepperell, MA 01437.
Leger, Eugene H. Head. Designer and builder of superinsulated houses.

W.M. Design Group, RFD 1, Box 123, Center Harbor, NH 03226.
Mead, William. Principal architect. Designer of double-envelope houses.

PERSONS WORKING INDEPENDENTLY

Arnow, Joshua, 14 Butler Rd., Scarsdale, NY 10583. (914) 725-1168. Builder and owner of double-envelope house.

Beale, Galen, Loudon Ridge Rd., Loudon, NH. Owns double-envelope house.

Bentley, R.P., Box 786-T, Tupper Lake, NY 12986. Designer and builder of superinsulated houses.

Burns, Warren, Boscawen, NH. Owns double-envelope house.

Conforti, Victor, 755 Broadway, Sonoma, CA 95476. Employs convective loop airflow serving greenhouse and windows.

Demmel, Dennis, Rt., 1, Box 234 B, Hartington, NE 68739. Has double-envelope house.

Dunn, Carroll (Rusty), Deerfield Dr., Lot 10, PO Box 1792, Truckee, CA 95734. Built several double-envelope houses.

Hart, John, Bridgewater, NH. Has double-envelope house.

Huber, Hank, PO Box 165, New Ipswich, NH 03071. Built several double-envelope houses.

Laz, Daniel, 4001 Farmington Dr., Champaign, IL. Owns superinsulated house.

Mastin, Robert, 1355 Green End Ave., Middletown, RI 02840. (401) 847-1488. Owns double-envelope house.

Phelps, Richard A., RR#1, Lerna, IL 62440.

Roggasch, Bob, 215 Ina St., Fairbanks, AK 99701. Or: Henderson Rd., Fairbank AK 99701. Built and owns superinsulated house.

Saunders, Norman B., 15 Ellis Rd., Weston, MA 02193. Made performance study of double-envelope house.

Weller, Richard D., 1407 Joseph St., Macomb, IL 61455. Owns a super-insulated house.

Zumfelde, Dale R., 11 Larkspur Lane, Batavia, OH 45103. Designed double-envelope house.

Appendix 2
GLOBAL INDEX OF INDIVIDUALS

Appendix 3

"SANTA CLAUS METHOD" OF SUPPLYING SUPPLEMENTARY HEAT

A radically new method of supplying supplementary solar heat to super-insulated houses-with-basement may convince owners of such houses that there really is a Santa Claus. The proposed method, designed to provide a moderate thermal boost and large carrythrough, employs (1) an enormous thermal mass --the basement itself-- maintained at 60 to 75F, (2) a high-COP air-conditioner to upgrade and transfer heat from basement to rooms, and (3) a small, cheap, air-type solar collector to replenish the basement's heat on sunny days.

(1) Storage: A typical concrete basement of a superinsulated house is well insulated and has a thermal mass of about 60 tons and a thermal capacity of 25,000 Btu°F. In cooling 10 F degrees it gives off 250,000 Btu -- enough to supply all the supplementary heat needed by a superinsulated house throughout about four overcast cold days in mid-winter. The thermal mass is about 5 times that of the house proper. Because the basement temperature is kept within the range 60 to 75F, the basement can be used in normal manner: used as workroom, playroom, bedroom, etc. -- while serving simultaneously as a thermal mass. Thus it is, in effect, a cost-free storage system.

(2) Transfer of heat to rooms: When, during cold nights or cold and overcast days in mid-winter, the rooms threaten to become cold, a modern, small, high-efficiency, air-conditioner is turned on by a thermostat (or by hand) and goes to work upgrading the basement heat (to about 80F) and delivering it to the rooms. The device is situated wholly indoors between basement and main hallway: it uses the basement as heat source and uses the hallway as heat sink; that is, it extracts heat from the basement air and discharges heat to the hallway. Some modern air-conditioners (for example, the G.E. AGDE 910F, 10,000 Btu/hr, 7½ amp., 115 v., plug-in air-conditioner) have a Coefficient of Performance (COP) of 3 when operated normally, i.e., to remove heat from a house, and have a COP of about 4 when used to deliver heat to a living area. (Some persons may not realize that an air-conditioner, when used to deliver heat, has higher COP than when used to remove heat. When the device is used to remove heat, the energy used to run the compressor and fans provides no direct thermal benefit; but when the device is used to deliver heat, such energy is straightforwardly helpful.) A COP of 4 is considerably higher than the COP of most heat-pumps! Yet the air-conditioner costs only a small fraction as much as a heat-pump, is easier to install, and avoids the grave problems that beset air-to-air heat-pumps when the outdoor temperature drops far below 32F. Also, the air-conditioner can be relocated in summer and put to work keeping the house cool. If a given superinsulated house would normally require about $120 worth of electric back-up heat per winter, it would require only $30 worth of electric power if the proposed scheme were used. The amount of electricity required would be so small as to make only a trivial addition to the utility company's load, and might not increase the peak load at all.

(3) Replenishing the basement heat: On sunny days in mid-winter a small (90 sq. ft.) air-type solar collector, mounted on the vertical south wall of the garage, would operate and would deliver (via duct and 100-watt fan) air at about 80F to the basement. Because this discharge temperature is so low, the collector could be of very simple type: single glazed and employing only 2 in. of insulation. The collector might consist merely of a flexible plastic assembly which, in summer, would be rolled up and stored in the garage. Such device might cost only $700.

Discussion

All three components are simple, cheap; each seems almost like a gift from Santa Claus. Even if the house occupants decide to use a higher-than-recommended rate of air-change, even if they switch to low-energy refrigerator and light bulbs, and even if the architect has skimped by having no vestibule-type doors, the proposed scheme should be ample to keep the rooms at about 70F throughout the winter. Even if the house has no furnace or stove or other auxiliary system, the occupants need have no qualms; and by having no furnace or stove, no chimney, no oil tank, they may save $2000 to $5000 and at the same time avoid the fire hazards and smells associated with furnaces and stoves and avoid the associated maintenance chores.

About half the cost of the air-conditioner can be charged to its use in summer.

I assume that an air-conditioner designed to operate in the range 70 to 120F can operate equally well in the range 60 to 75F. If I am wrong, then I would hope that the manufacturers can produce modified models meant for this lower-temperature range. I hope that COP even higher than 4 can be achieved, inasmuch as the proposed temperature step (about 10 or 20 F degrees) is much smaller than that which pertains to air-conditioners used in normal manner.

Why not use the solar collector to heat the basement somewhat hotter, say to 80 or 90F, so that there would no longer be a need for heat-upgrading, and a simple blower could be used instead of an air-conditioner? Because (a) the basement would then be too hot for normal occupancy and (b) the collector efficiency would be lower.

Why not install an 8-ton bin-of-stones and heat it to, say, 120F? Because this would add $1000 to the cost (i.e., an annual charge of about $120, assuming a 12% interest rate), preempt much space, reduce the efficiency of the collector, and store only about 1/3 as much heat.

Why not use the collector to heat the rooms directly? Because, during any sunny day, they are already warm enough -- thanks to the direct passive solar heating via the south windows.

Why not constructively link the solar domestic hot water (DHW) heating system (if any) and the air-to-air heat-exchanger (if any) to the above described system? Why not integrate them in cost-effective way? Good idea. The collector could contribute to preheating of DHW and the warm basement could contribute to the air-change process by making up for the losses normally associated with such process.

In summer, the occupants could operate the air-conditioner in such a way that it would cool the basement and would run only in the coldest (after midnight) hours. Thus its efficiency would be especially high and it would contribute nothing to the utility's peak load. During the hot part of the day, cool basement air could be circulated to the rooms by a small fan. (H.E. Thomason has for many years made excellent use of an air-conditioner operated just at night. The conditioner is a very small one, yet it provides cooling for the entire house; the total cost of operation per summer is very low.)

Appendix 4

NOTES ON THE FIFTH NATIONAL PASSIVE SOLAR CONFERENCE, OCTOBER 1980

Here I summarize the double-envelope house panel discussions and technical papers presented at the Fifth National Passive Solar Conference at Amherst, Mass., in October 1980. The conference as a whole was week-long. The portions dealing with double-envelope houses conformed to this schedule:

Monday evening 10/20/80, 5:30 p.m. - 8:30 p.m.
Tuesday morning 10/21/80, 6:30 a.m. - 7:30 a.m.
} Informal discussions among speakers

Tuesday 10/21/80, 8:30 a.m. - 6:00 p.m. Formal workshop managed by Don Booth, Hank Huber, and W.R.L. Mead. Attendance: about 800.

Wednesday 10/22/80, 8:30 a.m. - 1:00 p.m., and 4:00 p.m. - 6:30 p.m. Technical papers presented by designers and engineers. Attendance: about 700.

SPEAKERS

Akridge, James M., College of Architecture, Georgia Institute of Technology, Atlanta, GA 30332.

Berk, Jim, Solar Clime Designs, Box 9955, Stanford, CA 94305.

Booth, Don, Community Builders, Shaker Rd., Canterbury, NH 03224 (Co-planner of workshop).

Brodhead, William, Buffalo Homes, RFD 1, Box 209, Riegelsville, PA 18077.

Chen, Dr. Bing, University of Nebraska, Engineering Rm. 235, Omaha, NE 68182.

Converse, Prof. Alvin O., College of Engineering, Thayer Hall, Dartmouth College, Hanover, NH 03755.

Cowlishaw, Rick, 3246 Squaw Valley Dr., Colorado Springs, CO 80918.

Dougald, Prof. Donald E., School of Architecture, University of Virginia, Charlottesville, VA 22901.

Ghaffari, Homer T., Brookhaven National Laboratory, Upton, NY 11973.

Grundy, Prof. Roy, College of duPage, Glen Ellyn, IL 60137.

Henninge, Robert, Wild Turkey Hollow, Rt. 1, Box 26A, Cutler, OH 45724.

Holmes, Douglas B., 4 John Wilson Lane, Lexington, MA 02173.

Huber, Hank, PO Box 165, New Ipswich, NH 03071 (Co-planner of workshop).

Jones, Ralph F., Brookhaven National Laboratory, Upton, NY 11973.

Kitzmann, Dr. Gerald A., Salgary Energy Corp., 7 Innes Ave., New Paltz, NY 12561.

Kohler, Joe, Total Environmental Action, Box 47, Harrisville, NH 03450.

Maeda, Bruce T., Earth Integral Corp., 2655 Portage Bay, Suite 5, Davis, CA 95616.

Mastin, Robert, 1355 Green End Ave., Middletown, RI 02840.

Mead, William R. L., of the W. M. Design Group, RFD 1, Box 123, Center Harbor, NH 03226 (Co-planner of workshop).

Nicholson, Nick, Ayer's Cliff Centre for Solar Research, Walker Rd., PO Box 344, Ayer's Cliff, Quebec, Canada JOB 1CO.

Ortega, Joseph K. E., SERI, Golden, CO 80401.

Ray, James A., Ener-Tech Inc., 1924 Burlewood Dr., St. Louis, MO 63141.

Reno, Vic, Contemporary Systems, Inc., Rt. 12, Walpole, NH 03608.

Saunders, Norman B., 15 Ellis Rd., Weston, MA 02193.

Shurcliff, William A., 19 Appleton St., Cambridge, MA 02138.

Smith, Robert O., Robert O. Smith and Associates, 55 Chester St., Newton, MA 02161.

Smith, Tom, Positive Technologies Corp., PO Box 2356, Olympic Village, CA 95730.

INTRODUCTION

Excitement ran high at this two-day session on design and performance of double-envelope houses. Never before had so many double-envelope-house experts been gathered together, never before had discussions on this topic been planned with greater care, and never before had such a large audience (about 800 by actual count) assembled to listen to -- and question -- the experts.

The excitement stemmed mainly from the massive collision between two dedicated and vocal groups: a group strongly enthusiastic about double-envelope houses (mainly younger designers and architects, and owners and occupants of such houses), and a group impressed with the limitations, shortcomings, and uncertainties of such houses (mainly older architects and engineers involved in detailed measurements of thermal performance). Each side scored major points.

There were no bombshells, few decisive new facts. Few persons changed their minds.

If there was any consensus, it was this: The available data on thermal performance is distressingly meager and, in many instances, unreliable. Much more data-taking and analysis are needed. Although most double-envelope houses provide much comfort and require little auxiliary heat, and are much liked by the occupants, the rate of circulation of air in the convection loop is lower than contemplated, the amount of heat delivered to (and later drawn from) the crawl-space earth is small relative to the amount delivered to (and later drawn from) higher-up portions of the building structure, considerable variations in temperature are sometimes experienced in cold sunless periods in winter and during hot sunny days in summer, and the amount of auxiliary heat needed in winter corresponds to burning about 1/2 to 2 cords of wood. The general designs used are far from final; much more needs to be known about proper area of south glazing, proper use of vapor barriers, how to increase the rate of convective-loop airflow to some optimum value, and how to encourage storage of more heat in the crawl-space earth or other thermal mass situated low down in the building.

There was no consensus on this $64 question: In the next few years, as designers learn more about thermal performance and explore ways of overcoming the limitations, will they find themselves retaining the key hallmark features of double-envelope design or will they abandon some of these features and, instead, adopt features characteristic of <u>other types</u> of solar heated buildings? Will they greatly reduce the area of south glass? Employ fan-and-duct to transport hot air to low-down thermal mass? Eliminate the within-north-wall airspace? Will the double-envelope design persist as a unique species, or will it blend with more conventional designs and lose much of its identity?

TECHNICAL TOPICS

Area Of South Glass

Many speakers said that, in many cases, the areas of greenhouse south glass are too large. Uppermost south glass (especially, upper sloping south glass) should be curtailed.

Speed Of Airflow In Convective Loop

Many speakers said that effort should be made to speed this airflow. Avoid use of obstructing north-south timbers? Round the corners where the airflow direction changes? Keep R-value of upper part of outer north wall low, to insure that the loop air there is greatly cooled and becomes much more dense? Several investigators found that airflow rates were often only a few inches per second at locations where rates many times higher might be desired. Sometimes, even when moderately fast flow was expected, the flow was so slow that the flow-rate sensors failed to respond.

Reversal Of Flow At Night

Some investigators found that on cold nights the direction of airflow reversed, as expected. Others failed to find any reversal. In some instances, small-loop flow was found, i.e., flow involving just the greenhouse and the crawl-space below with little or no flow within the north wall airspace.

Vapor Barriers

Designers were urged to give more attention to vapor barriers serving the north wall system and other regions. Severe moisture problems have arisen in several attached-greenhouse-type houses in which moisture dangers had been given inadequate attention.

Fire Hazard

Several speakers referred to the need to install, within the north wall airspace, automatically closing dampers, or smoke or heat detectors, or gypsum-board facings. Others pointed out that an open stairwell in any house presents a similar fire-spread hazard. There was mention of a fire in a double-envelope house somewhere in California: a fire that to some extent involved part of the convection loop.

Minimal Heat Storage In Crawl-Space Earth

Many speakers indicated that the amount of heat stored in the crawl-space earth was very small. The desirability of increasing the amount stored here was indicated. Often, only the few top inches of earth contributed significantly. However, even a modest output of heat from this earth on cold nights can be very helpful in preventing the greenhouse from becoming too cold (colder than about 50F or in some instances about 40F). There was a consensus that much heat is stored relatively high up, specifically in the wood, gypsum, etc., components of the upper part of north wall system, ceiling or roof system.

Merit Of Fan-and-Duct System

Several speakers spoke out strongly in favor of installing a small fan-and-duct system to take hot air from the uppermost part of the greenhouse and deliver it to a storage system in the crawl-space or basement. They said that fans and ducts are simple and inexpensive and will deliver hot air when you want it, where you want it, and with scarcely any loss of heat en route. One designer-builder installed 2100 hollow concrete blocks in the crawl-space and employed a blower to drive hot air through the hundred or more channels in the array of blocks; he found that the scheme worked excellently: a very large amount of heat was stored. Two other speakers proposed using a fan to drive hot air (from the greenhouse) through a bin-of-stones beneath the main-story floor. If a fan is to be used, the designer can considerably reduce the width of the within-north-wall airspace, e.g., from 36 ft. to, say, a small fraction of this. Or, if a simple, circular-cross-section duct running directly downward from peak of greenhouse to the crawlspace is to be used, the air-space within the north wall can be eliminated entirely -- with consequent saving of space, simplification of north-wall and north-window construction, and elimination of danger of fire-spread within north wall. Some persons have a categorical dislike of dependence on electricity (as for driving fans); but it was pointed out that these persons are already using electricity for electric lights, TV, etc., and in any event on-site production of electricity by photovoltaic cells may be cost-effective within a few years.

Diverting Descending Cold Air In Greenhouse

On a cold night much cold air descends, within the greenhouse, very close to the greenhouse south glass. In some houses this cold air flows across the surface of the greenhouse earth (and plants) before entering the crawl-space. One speaker urged that a special bypass route be provided: a slot that is very close to the greenhouse glass and extends all the way along it, allowing the descending cold air to flow directly into the crawl-space without chilling the greenhouse earth and plants.

Cooling Tubes

For cooling several existing double-envelope houses, deep-buried galvanized steel tubes, 1½ or 2 ft. in diameter and 50 to 100 ft. long, are provided. On hot days in summer the house occupant opens vents near the top of the house; hot air flows out here and is replaced, one hopes, by outdoor air drawn in via the underground tubes. Although some speakers reported successful use of such tubes, other speakers warned that, by mid-summer, the earth adjacent to the tubes may be considerably warmed up and accordingly the cooling effect may become quite small. Also, humidity and moisture deposition may pose problems. In some instances the direction of airflow in the tubes was wrong; either because of improper direction of slope of the tubes or because of ill-advised opening of doors or windows of the house, house air flowed <u>out</u> via the tubes and hot outdoor air flowed directly <u>into</u> the house. One speaker said that, if an underground cooling tube is to be provided, it should be equipped with a fan so that the direction and speed of airflow will be fully controllable.

Vent Size

To insure that the upper parts of house and greenhouse do not become intolerably hot on sunny summer days, the designer must specify large vents; vent area should be about 15 to 20% of the greenhouse floor area, i.e., of the order of 35 to 50 sq. ft., typically. Alternatively, a large exhaust fan may be used. The air-change-rate goal may be as high as one complete change of greenhouse air per minute (or per few minutes). Natural convection may provide such a rate if the upper and lower vents are properly sized.

Post-and-Beam Construction

One New Hampshire builder is now building post-and-beam-type double-envelope houses. Much use is made of factory-built, large-area, sandwich-type panels that have a rigid foam core. Assembly is simple and fast. Performance is excellent.

Cost

Several speakers implied that double-envelope houses cost 5 to 10% more than casually built conventional houses. Several speakers suggested that prospective buyers of such houses be warned that the greenhouse is far from free: it may typically add $10,000 to $15,000 to the cost.

Computer Programs

Several groups have developed computer programs that permit one to predict in some detail the performance of a double-envelope house of given dimensions, etc. Rates of airflow, rates of heat-flow and storage, room temperatures, and auxiliary heat requirements can be calculated. Such calculations help a designer improve his design. Among the groups that have developed such programs are Ralph F. Jones et al of the Brookhaven National Laboratory and Norman B. Saunders of 15 Ellis Rd., Weston, MA 02193.

Appendix 5
NEW OFFICIAL RECOMMENDATIONS ON FIRE SAFETY

In mid-1980 the Council of American Building Officials (CABO), on behalf of the Department of Energy (DOE), formulated and published a set of detailed recommendations concerning the structure, durability and safety of solar heating systems including those of double-envelope houses.* The recommendations are directed mainly to local agencies that prepare and promulgate local building codes. Presumably many of the recommendations will be adopted by the local agencies, hence will be of interest to designers of double-envelope houses.

Some excerpts (from pages 20 and 21) dealing with fire safety and emergency egress are presented below.

B-111.3 Fire Safety

a. Firestopping

Commentary: Building codes require all concealed draft openings both horizontal and vertical to be "firestopped" to delay the spread of fire and smoke within the concealed space. Collectors that are building components often use the convective air movement or "stack effect" as a mechanism for distributing heat throughout a building, thus requiring a completely cleared opening for the entire height (wall) or length (roof) of the collector. Obviously, these requirements conflict when a wall collector is more than one story in height or a roof collector extends over at least one support. Compliance with the following is considered equivalent:

---

* "Recommended Requirements to Code Officials for Solar Heating, Cooling and Hot Water Systems: Model Document for Code Officials on Solar Heating and Cooling of Building", prepared by Council of American Building Officials. June 1980. 48 pages. Available from DOE Office of Solar Applications for Buildings. Document DOE/CS/34281-01. Also available from U.S. Government Printing Office; Document 1980-0-620-309/216. No indication of price.

1. "Firestopping" may be omitted in concealed spaces in combustible construction when smoke activated fire dampers are installed to cut the vertical draft opening at each floor level in walls and at each support in roofs.

2. "Firestopping" may be omitted in concealed spaces in non-combustible construction when smoke activated fire dampers are installed at each supply and return air opening. All wall cavities (concealed spaces) in excess of 200 square inches shall meet the requirements of a vertical shaft as indicated in Appendix XB-1.

EXCEPTION: In exterior walls only, the exterior wall of the shaft need not meet the vertical shaft requirements.

B-111.4 Emergency Egress

Commentary: The exit requirements of the building code requires that every sleeping room in a dwelling unit be provided with at least one window or exterior door permitting emergency egress or rescue. In some solar energy systems it is desirable, because of the location of sleeping rooms, to eliminate all openings in the exterior bedroom walls. Compliance with the following is considered equivalent.

An exterior door or window permitting egress or rescue is not required in sleeping rooms of dwelling units providing the room and adjacent hallway has a listed smoke detector and the door from the room opens directly into a corridor that has access to two remote exits in opposite directions. Both smoke detectors shall be interconnected in a manner such than when one detector is activated the other detector is activated.

BIBLIOGRAPHY

A-256 ASHRAE, Program abstracts for conference on "Thermal Performance of the Exterior Envelopes of Buildings: at Kissimmee, FL, Dec. 3-5, 1979.

B-237 Bentley, R.P., "Thermal Efficiency Construction", publ. from Box 786-T, Tupper Lake, NY 12986. 1975. 96 p. $15.75.

B-250 Besant, R.W., R.S. Dumont, and Greg Schoenau, "The Saskatchewan Conservation House: Some Preliminary Performance Results", Energy and Buildings, 2, 163 (1979).

B-251 Besant, R.W., R.S. Dumont, and G.J. Schoenau, "The Passive Performance of the Saskatchewan Conservation House", paper presented at the AS-ISES Passive Solar Conference, San Jose, Jan. 1979.

B-251a Besant, R.W., R.S. Dumont, and D. Van Ee, "An Air to Air Heat Exchanger for Residences", an engineering bulletin published (in 1978) by the Extension Division, University of Saskatchewan, Saskatoon, Sask. Canada S7N 0W0. $2.

B-251c Besant, R.W., R.S. Dumont, and G.J. Schoenau, "The Saskatchewan Conservation House: A Year of Performance Data", a paper presented at the 1979 Annual Conference of the Solar Energy Society of Canada, Inc., at Charlottetown, P.E.I., Canada.

B-251e Besant, R.W., R.S. Dumont, and G.J. Schoenau, "A Survey of Some Recently Constructed Passive Solar Buildings in Saskatoon", a paper presented at the 1979 Conference at Charlottetown, P.E.I., Canada. $5.

B-415 Booth, Don, "The Double Shell Solar House", publ. by Community Builders, Shaker Rd., Canterbury, NH 03224. 1980. About 120 p. $15.75.

B-650 Bushnell, R.H. "Climatic Kelvin-Day Values Above or Below Any Base", Monthly Weather Review 107, 1083 (1979).

D-11 Dallaire, Gene, "Zero Energy House: Bold, Low-Cost Breakthrough That May Revolutionize Housing", Civil Eng'g., May 1980, p. 47 - 59.

G-180 Ghaffari, H.T., R.F. Jones, and G. Dennehy, "Double Shell House Measured Thermal Performance Robert and Elizabeth Mastin Ekose'a House Middletown, Rhode Island", Feb. 1980. 21 p. Published by Brookhaven National Laboratory, Upton, New York, 11973.

G-880 Gunnarshaug, Johs., "Low Energy Experimental Houses", a paper presented at a November, 1976, meeting in Brussels.

H-62h Harris, W.S., et al, "Energy Requirements for Residential Heating Systems", Report D/272 of July 1, 1965, from the University of Illinois. About 15 p. Discusses shortcomings of degree-day concept.

J-400 Jones, R.F., R.F. Krajewski, and Gerald Dennehy, "Case Study of the Brownell Low Energy Requirement House", published by Brookhaven Nat'l. Laboratory, May 1979. Report BNL 50968, UC-95d. TID 4500.

L-195 Lewis, Daniel, and Winslow Fuller (D. Lewis, W. Fuller), "Restraint in Sizing Direct Gain Systems", Solar Age, Dec. 1979, p. 28.

M-82a Massdesign Architects and Planners, Inc., "Solar Heated Houses for New England", April 1975. Published by the authors, 138 Mt. Auburn St., Cambridge, MA 02138.

M-92 McGrath, Edward, "How to Build a Superinsulated House", privately published. 1978. 37 p. $4. Write to Project 2020, Box 80707, College, AK 99708.

N-406 New England Solar Energy Assn., "Air Envelope House: Report on an Informal Technical Conference, July 21, 1979". About 25 p. $5. Summarizes 1979 Conference at Boscawen, N.H.

\- Popular Science, "Double-Shell Solar House: High Performance in a Controversial Package", by Ron Dans. Dec. 1979. Excellent article

R-11 Rahders, R.R., "Your House Can Do It: The Passive Approach to Free Heating and Cooling", published by Thacher & Thompson, 215 Oregon S Santa Cruz, CA 95060. 1979. 170 p. $12.

R-190 Robinson, D.A., "Insulating a Solar House", a 6-p. paper presented at the 1978 annual meeting of the AS-ISES. See Proceedings, p. 196.

R-192 Robinson, D.A., "Simple Modification of Current Building Practice Applied to the Construction of a Solar-Assisted Super-Insulated House", paper presented at 3rd National Passive Solar Conference, San Jose, Calif.

R-194 Robinson, D.A., "Conservation: The Other Half of Solar Design", paper presented at 4th National Passive Solar Conference, Kansas City, '7

R-196 Robinson, D.A., "The Art of the Possible in Home Insulation", Solar Age, Oct. 1979, p. 24.

R-205 Rodgers, Russell, "Performance of a Double Skinned Envelope House in Nelson, B.C.", publ. by One Step Ahead Energy Systems Ltd., 301 Vernon St., Nelson, B.C. Canada V1L 4E3. Describes Kootenay House #1. 1980.

R-206 Rodgers, Russell, "Kootenay House #1" Same date, same publisher. Gives additional performance data.

R-260 Rosenfeld, A.H., et al: "Building Energy Use Compilation and Analysis an International Comparison and Critical Review. Part A: New Residential Buildings", Report LBL-8912-rev., publ. by Lawrence Berkeley Laboratory of the University of California. 39 p. Aug. 1980. Presents performance data on several superinsulated houses.

S-184-C2.3 Shick, W.L., R.A. Jones, W.S. Harris, S. Konzo, "Circular C2.3, Illinois Lo-Cal House", published by Small Homes Council, Universit of Illinois, Urbana, IL 61801. Spring 1976. 8 p. 50¢ postpaid.

S-185 Shick, W.L., R.A. Jones, W.S. Harris, and S. Konzo, "Technical Note 14: Details and Engineering Analysis of the Illinois Lo-Cal House", published by Small Homes Council, Building Research Council, University of Illinois, Urbana, IL 61801. May 1979, 110 p. $8.50.

S-235aa Shurcliff, W.A., "Solar Heated Buildings of North America: 120 Outstanding Examples", Brick House Publishing Co., Andover, MA 01810. (1978). 300 p. $8.95 paperback, $14.95 hard cover.

S-235cc Shurcliff, W.A., "New Inventions in Low-Cost Solar Heating: 100 Daring Schemes Tried and Untried", Brick House Publishing Co., Andover, MA 01810. (1979). 300 p. $12.00 paperback.

S-235ee Shurcliff, W.A., "Thermal Shutters and Shades", Brick House Publishin Co., Andover, MA 01810. (1980). $12.95 paperback, $24.50 hard cove

S-290 Smith, Tom, and L.P. Butler, "Energy Producing House: Handbook Case Study of a Passive Solar House", published by Ekose'a, 573 Mission St., San Francisco, CA 94105. (1978) 55 p. $18.50.

\- Solar Clime Designs, "Convection Loops", a small periodical. Editor: Jim Berk. Cost: $7.50 per year. Write to Box 9955, Stanford, CA 94305.

U-915 University of Saskatchewan, Division of Extension and Community Relations, "Low Energy Passive Solar Housing Handbook", (1979). 38 p. Describes a variety of insulation procedures such as those used in the Saskatchewan Conservation House.

W-120 Weber, D.D., W.O. Wray, and R.J. Kearney, "Simultaneous Modeling of Interzone Heat-Transfer by Natural Convection", a paper presented at the Oct. 1979 4'th National Passive Solar Conference in Kansas City. See Proceedings p. 231.

U-520-12 University of California, Lawrence Berkeley Laboratory, "Use of Mechanical Ventilation with Heat Recovery for Controlling Radon and Radon-Daughter Concentrations", by W.W. Nazaroff et al, LBL-10222, March 1980. 22 p.

U-520-13 University of California, Lawrence Berkeley Laboratory, "Residential Ventilation with Heat Recovery: Improving Indoor Air Quality and Saving Energy", G.D. Roseme et al, LBL-9749, 1980. 29 p.

INDEX